A FIELD GUIDE TO BERRIES AND BERRYLIKE FRUITS

A FIELD GUIDE TO

BERRIES

AND BERRYLIKE FRUITS

BY MADELINE ANGELL

ILLUSTRATIONS BY MARIE SUILMANN

THE BOBBS-MERRILL COMPANY, INC. INDIANAPOLIS/NEW YORK

The author wishes to thank Bruce Ause, Director of the
Environmental Learning Center in Red Wing, Minnesota, for
answering several of her questions. She also wishes to thank Dr.
Richard Spellenberg, Professor of Biology, New Mexico State
University, for help in classifying Allthorn, and Dick Behrens for
locating several specimens.

Library of Congress Cataloging in Publication Data

Angell, Madeline.
  A field guide to berries and berrylike
fruits.

  1. Berries — United States — Identification.
2. Berries — Canada — Identification. 3. Wild
plants, Edible — United States — Identification.
4. Wild plants, Edible — Canada — Identification.
I. Suilmann, Marie. II. Title.
QK660.A58   582'.0464   80-2730
ISBN 0-672-52676-X   AACR2
ISBN 0-672-52695-6 paper

Designed by Jean Callan King
Manufactured in the United States of America

First printing

CONTENTS

To Dorothy Whitmer,
a very welcome addition to the family circle;
and to my biking friends.

INTRODUCTION

A field guide for the identification of wild berries and berrylike fruits has long been needed. Information on berries is contained in numerous books on wildflowers, shrubs and trees, and there are books available on edible wild plants, but a book on wild berries has been lacking. Most people who enjoy getting out in the country to hike, bike, bird-watch or hunt have the same desire to identify a berry, and learn something about it, as they do to become knowledgeable about other aspects of the outdoors. Even the cross-country skier, observing a cluster of withered berries hanging on a shrub or vine, may be curious as to the identity of those berries.

Since this book is intended for the average nature-lover, it includes not only fruits that are technically berries, but those that look like berries and those we commonly think of as berries. A book including only fruits that are technically berries would have to leave out raspberries, blackberries, strawberries and many others with the word "berry" in their names. Raspberries and blackberries are not berries, but clusters of tiny drupes, each of which has the same structure as a plum. A strawberry is not a berry but a cluster of achenes (seeds) contained in pits on the surface of an enlarged, fleshy receptacle. Other fruits that we are *not* apt to think of as berries, such as pawpaw, persimmon and cactus fruits, are actually berries and are therefore included. Tomatoes, grapefruits and bananas are also berries, but are not included here because they do not grow wild in North America.

There is not complete agreement among authorities as to the definition of a berry. Like other fruits, a berry is a developing or ripened ovary, sometimes including other adherent parts. Some botanists insist that a berry must have two or more seeds, thus distinguishing it from a drupe, which has only one. Others define the difference between berry and drupe by the fact that the drupe's one seed is enclosed in a hard and bony covering, a pit or stone, such as that of a plum. The definition followed in this book is that a berry is a simple fruit (consisting of one piece) which is fleshy or pulpy throughout, with seeds imbedded in the pulp. Blueberries, cranberries, currants, gooseberries, grapes and honeysuckle fruits are typical berries.

Plums and cherries are typical drupes. However, many of them are included here because they look like berries. Some berrylike fruits are pomes. The name "Juneberry" reveals that it is commonly regarded as a berry. Actually, it is a pome, a fleshy fruit with a core. The pome we are most familiar with is the apple. Drawing the line between a drupe or pome which is berrylike and one which is not is relative. Any fruits that are over an inch in diameter are not included unless such fruits are technically berries. Among wild apples, only the Oregon Crab Apple has been included, because it is ¾ of an inch or less in diameter and it looks somewhat like a berry.

Berries serve an important function in reproduction of a plant. Like other fruits, they are usually green or white at first, remaining inconspicuous until they are ripe and the seeds are well formed, when flavor and texture are at their best. In temperate regions, birds are the principal consumers of these berries, because berries on high bushes or trees are inaccessible to animals that do not climb, and can only be consumed by them when they fall to the ground. The seeds are well protected by tough seed coats, and they are either discarded by the creatures who eat them or passed through the digestive systems unharmed, often at a considerable distance from the parent plant. In this way, the seeds are dispersed, spreading the species to other areas.

Many berries are prized by man because of their fine flavor and nutritive value. Some are mildly poisonous, a few are dangerously poisonous. Some berries are edible but unpalatable and are regarded only as survival food. In certain cases, the edibility of the berry is still unknown. Facts about a berry's edibility and suggestions for its use are given for each berry. Information on edibility is based primarily on *Human Poisoning from Native and Cultivated Plants* by James W. Hardin and Jay M. Arena, M.D., and on *A Field Guide to Edible Wild Plants*, by Lee Peterson. It is good to remember that birds and animals can often safely eat berries that are poisonous to humans. Berries that are only mildly poisonous to adults may be fatal to a child. When in doubt about the edibility of a berry, don't eat it. Even if you are quite sure of the identity of a berry and it is listed as edible, sample only a few berries the first time. An allergic reaction could cause great difficulty.

For ease of identification, the berries and berrylike fruits in this book are arranged according to the color of the mature berry — red, blue, purple, white or green, black, yellow or orange. Since most berries start out green or white, only those that are green or white at maturity are in the white-green section. Many blue and purple berries go through a red stage, and in such cases cross references will be found. Berries that are black at maturity may be red, blue or purple when immature, and here again cross references are made.

There is variation even in the color of the mature berry. The presence or absence of bloom (a whitish powder) on a berry affects its color. A black berry may look blue or purple if it is coated with a heavy bloom. Berry color is somewhat influenced by the way the light hits it or by the eye of the observer. A berry that looks purple to one person may look blue to a second person and purplish black to a third individual. If the berry you are trying to identify is purplish and you do not find it in the purple section, try the red, the blue or the black section.

In each color section, plants are divided into (1) flowering herbs, (2) woody vines and shrubs and (3) trees. The demarcation between shrubs and trees is an indefinite one, since a shrub is defined as a woody perennial smaller than a tree. Many of the plants described in the following pages are included in books on shrubs and also in books on trees. The deciding factor was whether the plant is most often seen in the shrub form or the tree form. The plants in each category are arranged according to families.

Some berry names are used for a number of different species. "Partridgeberry" is one such term: it is used for numerous plants, including cranberries and wintergreen. Only in the case of *Mitchella repens* is it used correctly. Presumably the reason for this name being used so often is that the partridge (a name often used erroneously for the ruffed grouse) is fond of the berries. In most cases, a popular name that would only cause confusion has been omitted.

Although an attempt has been made to include berries from all parts of the United States, a book of this size cannot claim to be exhaustive. When a berry grows in just part of one state or when it is very rare in North America, it is usually not included.

Scientific terminology has been avoided as much as possible. A glossary is included for explanation of technical terms found necessary.

There are more berries in the red section than in any other section. The color ranges from pink to purplish red.

FLOWERING HERBS • ARUM FAMILY

JACK-IN-THE-PULPIT
Arisaema triphyllum

Other names: INDIAN TURNIP, INDIAN JACK-IN-THE-PULPIT

Description: Unmistakable in appearance, this perennial has a flaplike specialized leaf called a spathe that is usually striped in shades of green and purplish brown. This 2- to 4-inch flap folds gracefully over the flowering part of the plant, the spadix, a clublike structure with small, greenish white flowers crowded near its base. The spadix is the "jack" and the funnel-shaped spathe that folds over it is the "pulpit." The plant grows 1 to 2 feet tall. It has 1 or 2 long-stalked, compound leaves, each with 3 pointed leaflets. During the summer the flowers are replaced by an egg-shaped cluster of glossy, round berries. They are green at first, but turn *bright red* with maturity, in July or August. As the spathe withers, the berry cluster emerges from its hiding place and becomes very conspicuous.

Habitat: This unusual plant grows in woods and swampy areas. Some authorities divide Jack-in-the-Pulpit into three species: this one; Woodland Jack-in-the-Pulpit (*A. atrorubens*); and Indian Turnip or Northern Jack-in-the-Pulpit (*A. stewardsonii*). One or another of these species is found from New England to Minnesota and south to Florida and Louisiana.

Remarks: The thickened underground stem (corm) of this plant contains crystals of calcium oxalate. These cause an intense burning sensation if the corm is eaten raw. The Indians learned that drying the corms made them palatable. They ate large quantities of them and used them for medicinal purposes to treat coughs and tuberculosis, among other things. The properly prepared corms are still enjoyed by people who relish wild foods. They can be cut into very thin slices, thoroughly dried, roasted and then eaten like potato chips. Boiling is no substitute for thorough drying.

Berry use: The berries are edible but peppery, and should be used with

Jack-in-the-Pulpit

great caution. Thrushes, pheasants and wild turkeys eat them, as do squirrels and mice.

Similar species: GREEN DRAGON (*A. dracontium*), also called DRAGON ARUM and DRAGONROOT, is less conspicuous than Jack-in-the-Pulpit, which it resembles. Its solitary leaves are divided into between 5 and 17 segments. It can be distinguished from Jack-in-the-Pulpit by the long tip of its narrow spadix, which extends far beyond the spathe. It is found from Quebec to Wisconsin and south, though it is nowhere abundant. Its mature berries are *orangish red or reddish orange*. It is best not to eat them.

WILD CALLA
Calla palustris

Other names: WATER-ARUM

Description: This is an attractive swamp flower, growing 5 to 10 inches tall. Its glossy leaves are heart-shaped, tapering to a point. Like Jack-in-the-Pulpit, it has a flaplike specialized leaf called a spathe, but in the case of Wild Calla the spathe is white. The flowering part of the plant, the spadix, is yellow and is covered with tiny, greenish white flowers. In mid- or late summer the flowers are replaced by a cluster of *red* berries.

Habitat: Wild Calla is found in bogs and swamps and along the edges of ponds from Quebec to Alberta, south to Virginia, Texas and Colorado.

Remarks: The fertilization of this plant is unusual: it is fertilized by pond snails. Thoroughly dried seeds and rootstocks can be ground into a flour that is nutritious but not very tasty.

Berry use: Eating any part of the raw plant causes an intense burning sensation. However, the thoroughly dried berries are edible.

FLOWERING HERBS • LILY FAMILY

ASPARAGUS
Asparagus officinalis

Description: We are most familiar with the young shoots of this plant, which are thick and become tender when cooked. At the tips are the brownish, scalelike leaves. As the plant matures, the stalk becomes

Wild Calla

tough; numerous tall, feathery branches, like those of a delicate fern, grow from the leaf axils. At this stage the branchlets add grace and beauty to a bouquet. The plant grows to 6 feet. Tiny, greenish yellow flowers dangle from side branches during May and June. The glistening berries are small, round and *red.*

Habitat: This is a garden escapee that grows in loose, sandy loam. It is found in eastern North America.

Remarks: Asparagus has been cultivated as a garden vegetable for over 2,000 years. It was a favorite with the ancient Romans.

Berry use: Although the berries are not commonly used, they can be crushed and used as flavoring for soups or wild meats. Great caution is advised.

FALSE SOLOMON'S-SEAL

Smilacina racemosa

Other names: FALSE SPIKENARD, SOLOMON'S-PLUME, SOLOMON'S-ZIGZAG, BRANCHED SOLOMON'S-SEAL

Description: This plant has an arching, sometimes zigzagging stem 1 to 3 feet tall. Its leaves are oval, pointed, alternately arranged and hairy underneath. Scars of former stems mark the fleshy rootstocks with circular "seals." Small, creamy white flowers grow in large, dense, frothy terminal clusters. The small, aromatic berry is at first white or green, speckled with brown or gold, but it becomes mottled *ruby red* when it matures in late summer.

Habitat: False Solomon's-Seal grows in dry to moist woods and thickets from Nova Scotia to British Columbia, south to Georgia, Missouri, Arizona, and in the mountains of the Pacific states.

Remarks: False Solomon's-Seal can be distinguished from Solomon's-Seal by the fact that true Solomon's-Seal has its flowers in the leaf axils rather than in terminal clusters. The young shoots of False Solomon's-Seal can be used in salads or cooked like asparagus. The rootstocks are also edible if soaked overnight in lye and parboiled.

Berry use: The berries are edible but not very tasty, and since they have a mild cathartic effect, caution should be observed.

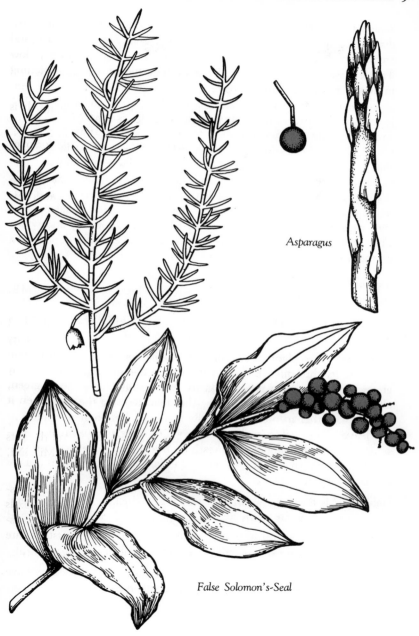

Asparagus

False Solomon's-Seal

Similar species: THREE-LEAVED FALSE SOLOMON'S-SEAL (*S. trifolia*) can be distinguished from False Solomon's-Seal by the fact that it has a short, erect stem about 6 inches in length and seldom has more than 3 leaves. The leaf bases taper and clasp the stem. The berries are *deep red,* and information on their edibility is lacking. This plant grows in wet woods and bogs from northern Canada south to Minnesota and west to New Jersey and Connecticut. STARRY FALSE SOLOMON'S-SEAL (*S. stellata*) is also called STAR-FLOWERED SOLOMON'S-SEAL and STARRY SOLOMON-PLUME. It can be distinguished from False Solomon's-Seal by the fact that it is smaller, with larger, starry flowers, fewer in number. The leaves clasp the zigzag stem and are close together. The berries are *at first green striped with black,* but with maturity they become *dull red, almost black.* They are edible but are full of seeds and should be used with caution. There are reports of their being used for jelly and syrup. This species grows in moist, open places, thickets and dunes from southern Canada south to Virginia and California.

TWISTED-STALK
Streptopus amplexifolius

Other names: WHITE MANDARIN, CLASPING-LEAVED TWISTED-STALK, LIVERBERRY, SCOOT BERRY

Description: This plant can be distinguished from similar plants by the fact that its threadlike flower stalks have a definite kink in them. It grows 3 feet tall with a zigzag stalk. Its oval, pointed leaves are arranged alternately at regular intervals along the stem, clasping the stem at their bases. Like other members of the lily family, it has parallel veining in the leaves. The bell-shaped, greenish white flowers grow singly (occasionally in pairs), one flower beneath each leaf, and have recurved petals. The berries, oval in shape, turn *red* in late summer. There is usually just one berry to a plant.

Habitat: This flower grows in cool woods and thickets from Greenland to Alaska and as far south as the mountains of North Carolina and Arizona.

Remarks: The young shoots can be added to salads for a cucumberlike flavor.

Berry use: The berries can be eaten in small quantities or made into

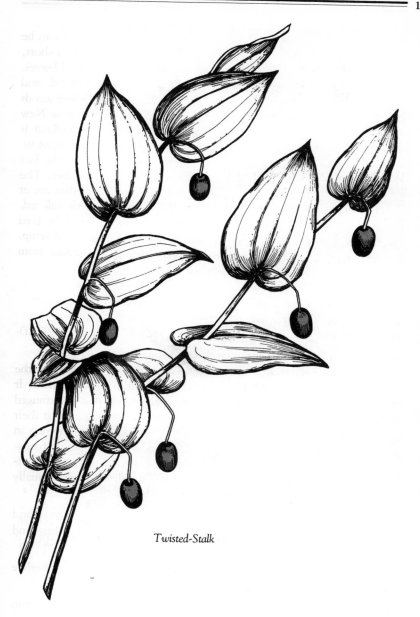

Twisted-Stalk

jelly, but they have a mild laxative effect and should be used with caution.

Similar species: ROSE TWISTED STALK (*S. roseus*), also called SESSILE-LEAVED TWISTED-STALK, is similar to the above, but it has more flowers and more berries. The flowers are rose or purple. The translucent red berries are very conspicuous in the fall. This species grows from southern Labrador to Alaska and south to the mountains of Georgia and Arizona.

WILD LILY-OF-THE-VALLEY
Maianthemum canadense

Other names: TWO-LEAVED SOLOMON'S-SEAL, CANADA MAY-FLOWER

Description: This is a low perennial, usually 3 to 6 inches tall. It has a slender, erect, often zigzag stem normally bearing 2 leaves but occasionally 1 or 3. The leaves are waxy and smooth, oval, pointed, heart-shaped near the base. The 4-pointed, white flowers grow in small, dense, upright clusters 1 to 2 inches long. The small berries are at first white and speckled, but when mature, in early fall, they are a pale, translucent red.

Habitat: Wild Lily-of-the-Valley grows in woods and forests from Newfoundland to Alaska, south to northern California, Iowa, and the uplands of Georgia and North Carolina. It does well in heavy shade and often makes a pleasant ground cover over large areas. It is a common woodland flower.

Berry use: Although one nature book states that these berries are edible, they should be used with great caution.

Similar species: LILY-OF-THE-VALLEY (*Convallaria majalis*) is a familiar flower because it is so often cultivated. Its white, bell-shaped flowers appear in May or early June. The red berry, which seldom forms, is *poisonous.* Animals usually leave the plant alone, but cases have been reported of their being poisoned by eating it. Lily-of-the-Valley is native to the Alleghenies, from Virginia to Tennessee.

Wild Lily-of-the-Valley

Lily-of-the-Valley

TRILLIUMS
Trillium spp.

Trilliums are such lovely spring flowers that they have been over-picked, and some states now have laws against picking them. Three leaves are whorled about the stem; the large solitary flowers have 3 petals and 3 sepals. There are numerous species of trillium, including several western ones. Those with red berries include:

PAINTED TRILLIUM
Trillium undulatum

Other names: PAINTED WAKEROBIN

Description: This trillium is outstandingly lovely, even in comparison with its beautiful relatives. It has a slender stem, 8 to 20 inches high, bearing egg-shaped, waxy green leaves that taper to a sharp point. Large, white petals, gracefully recurved and wavy-edged, extend beyond the sepals. At the heart of the flower is a scarlet or magenta blaze of color. The berries are oval, bright shining *scarlet.*

Habitat: This species is found in cold, moist woodlands, from Quebec and Ontario to Wisconsin, south to Georgia and Missouri.

Berry use: The berries are inedible.

Similar species: WESTERN WAKEROBIN (*T. ovatum*) grows in the North-west. RED TRILLIUM (*T. erectum*), also called WAKEROBIN, BIRTHROOT, and STINKING BENJAMIN, has the same range. Its flowers, which have a foul odor, are maroon or purple, from 2 to 3 inches wide. The berries are inedible.

FAIRYBELLS
Disporum lanuginosum

Other names: YELLOW MANDARIN, HAIRY DISPORUM

Description: Fairybells have forked stems bearing several stalkless, ob-long, alternate leaves, the last 2 being opposite each other. Nodding greenish-white, bell-shaped flowers yield smooth, *red,* egg-shaped berries.

Habitat: Fairybells like rich woods and mountain areas from Ontario, Ohio and New York south to Georgia and Alabama.

Berry use: Wartberry Fairybells (see below) are sweet and were eaten raw by Blackfoot Indians, but caution is advised for all fairybells.

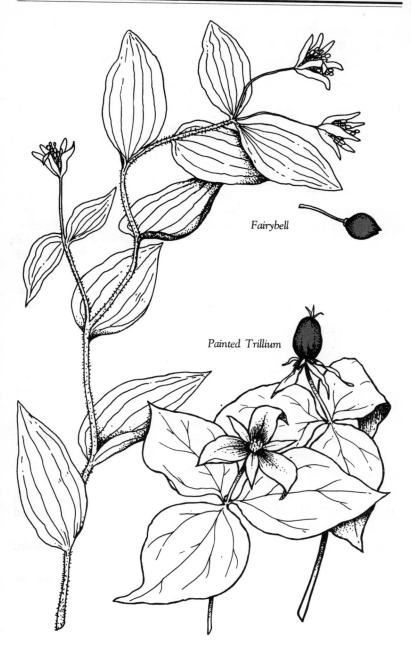

Fairybell

Painted Trillium

Similar species: WARTBERRY FAIRYBELL (*D. trachycarpum*) is often confused with Twisted-Stalk and False Solomon's-Seal. Flowers and berries are at the ends of the branches. The berries are *yellow or orange, becoming red when mature* in June. This species grows in wooded areas from British Columbia to Oregon, east to North Dakota, south through the Rocky Mountain region to Arizona and New Mexico.

FLOWERING HERBS • SANDALWOOD FAMILY

NORTHERN COMANDRA
Comandra livida

Other names: GEOCAULON

Description: This is a small plant with an erect, non-branching stem, from 4 to 12 inches tall. The leaves, which sometimes have a purplish tinge, are thin, alternate, elliptic, with short leafstalks. The flowers bloom on long stalks. What appear to be petals are really sepals, bronze or greenish white, triangular in shape. The berrylike drupes are *scarlet.*

Habitat: Northern Comandra likes moss or damp humus and is usually found at higher altitudes in New England, west to British Columbia.

Use of berrylike fruit: The fruit is fleshy and edible, but caution is advised.

FLOWERING HERBS • GOOSEFOOT FAMILY

STRAWBERRY-BLITE
Chenopodium capitatum

Description: This plant, in the same family as spinach and beets, has weak, drooping stems growing 6 to 24 inches tall. Triangular, irregular leaves have undulating or roughly toothed edges. Small greenish flowers mature into *bright red* fruits somewhat resembling raspberies. They cluster along the stalk and at the tip.

Habitat: Strawberry-Blite grows in sunny, disturbed ground, such as newly burned clearings in Canada and the northern United States.

Use of berrylike fruit: The fruits are edible, raw or cooked. They are nutritious but somewhat insipid.

POKEWEED
(Pokeweed Family) See Purple Berries.

Northern Comandra

Strawberry-Blite

FLOWERING HERBS • BUTTERCUP FAMILY

RED BANEBERRY
Actaea rubra

Also called Black Cohosh, this plant has *poisonous berries*. See White Baneberry in White Berries for a description. The main differences between these two species are that Red Baneberry has a rounder flower cluster, its berries are a *shiny red* and they are on thinner stalks. Red Baneberry is found from Labrador to Alaska, south to New Jersey, Indiana, Iowa, and throughout the western states.

GOLDENSEAL
Hydrastis canadensis

Other names: ORANGEROOT, GROUND-RASPBERRY, EYEBALM, EYEBRIGHT, EYEROOT, INDIAN DYE, INDIAN PAINT

Description: This stocky perennial has a thick, yellow rhizome that sends up one heavily veined basal leaf and a hairy stem with two small leaves near the top. The inconspicuous greenish white flowers have no petals, and the 3 sepals fall out when they open. The fruit is a head of tiny *orangeish red* berries that cluster together and resemble a raspberry.

Habitat: This plant likes rich, shady woods from Vermont to Virginia, west to Minnesota and Arkansas.

Remarks: The rhizomes of Goldenseal have medicinal value in the form of the drug hydrastine. Like Ginseng, it has been overcollected and is now very rare. Gardeners can help save the species from extinction by buying plants from nurseries selling wildflowers.

Berry use: The berries are *poisonous*. The plant contains alkaloids which cause ulcerations and inflammation of mucous surfaces.

FLOWERING HERBS • ROSE FAMILY

CLOUDBERRY
See Yellow or Orange Berries.

COMMON STRAWBERRY
Fragaria virginiana

Other names: WILD STRAWBERRY, SCARLET STRAWBERRY, HEART BERRY

Red Baneberry

Goldenseal

Description: This plant is similar to the domestic strawberry, but the "berries" are smaller. It grows close to the ground and sends out horizontal runners. There are usually several plants growing close together. The stems and leaflets are hairy. Three coarsely toothed leaflets grow on a slender stalk. The flowers, with their 5 round, white petals, grow on a separate stalk not longer than the leafstalk. The pulpy red (*occasionally white*) fruits hide under the leaves. They are so well known that they need no description. Although wild strawberries are smaller than domestic varieties, they are sweeter and more flavorful. Technically, what we call a strawberry is not a berry but the enlarged flower receptacle containing the real fruits, dry, hard seeds imbedded in pits on the surface. Strawberries ripen in early summer.

Habitat: Strawberries of one kind or another grow in meadows, fields, woods and forest edges throughout the country, with the exception of very dry areas.

Remarks: The name "fragaria" comes from the Latin word for fragrant and describes the odor of "berries" ready to be eaten. Early colonists made tea from strawberry leaves, which are rich in vitamin C, for prevention or treatment of scurvy. Such a tea was also used at one time to treat diarrhea.

Use of berrylike fruit: Wild strawberries are delicious eaten raw, or they can be used in any recipe calling for cultivated strawberries. They are rich in vitamin C. Indians ate strawberries when they had colds, and used the fruits for flavoring stews. Deer enjoy the berries, as do robins, bluebirds and downy woodpeckers.

Similar species: WOOD STRAWBERRY (*F. vesca*) is much like Common Strawberry except that the flowers and fruits are smaller and the fruit is usually held above the leaves. The edible fruit is more cone-shaped and the seeds are on the surface rather than in pits. BEACH STRAWBERRY (*F. chiloensis*) is planted as a ground cover and has edible tasteless fruits. It is found on coastal sand dunes and beaches from Alaska to South America. INDIAN STRAWBERRY (*Duchesnea indica*) resembles Common Strawberry, but has yellow flowers with 3-toothed bracts that extend beyond the petals and sepals. The fruit looks like that of the Common Strawberry but is inedible.

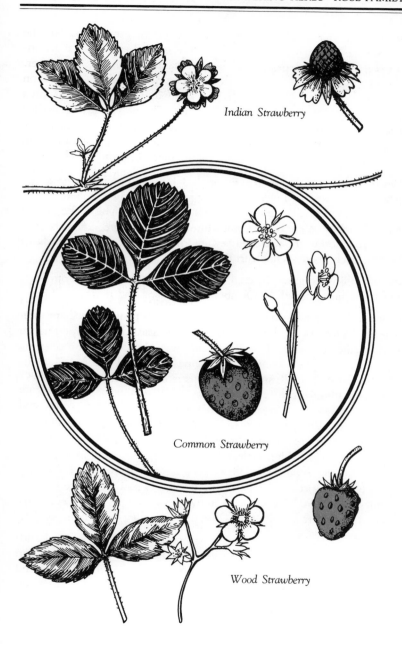

Indian Strawberry

Common Strawberry

Wood Strawberry

FLOWERING HERBS • GINSENG FAMILY

AMERICAN SPIKENARD
See Purple Berries.

GINSENG
Panax quinquefolius

Other names: AMERICAN GINSENG, MANROOT, MAN'S HEALTH, REDBERRY

Description: On a low stem, Ginseng's 3 long-stalked leaves grow in a whorl. Each leaf is divided into 5 spreading, toothed leaflets. The 2 lower ones are smaller than the others. The total height of the plant is 8 to 16 inches. A cluster of small white or yellowish green flowers grows at the junction of the leaves. They are followed by a cluster of *bright-red* berries, each about ¼ inch in diameter.

Habitat: Ginseng grows in cool, rich soil in shaded forests from Manitoba to Quebec, south to Oklahoma, Louisiana and northern Florida.

Remarks: The spindle-shaped, forked root of this plant has a shape often resembling that of the human figure and is regarded by many as having great medicinal value, as well as being an aphrodisiac. The scientific name "Panax" has reference to its being considered a cure for all ills. Modern research in Russia has demonstrated that the root serves as a tonic and stimulant.

Ginseng was dug in great quantities by early settlers, and as a result it is now rare. When the Indians gathered the roots, they usually bent down the stem and covered it and the ripened berries for continued propagation. The white settlers did not follow this good example.

It is sometimes cultivated, but the price per pound is less for cultivated plants. Wisconsin, one of the leading Ginseng producers in this country, sends 90% of its market overseas. The roots are collected, dried and exported, mainly to China, where they have been used for thousands of years.

Berry use: The berries can be eaten raw but are not good. See also Dwarf Ginseng, Yellow or Orange Berries.

Ginseng

FLOWERING HERBS • DOGWOOD FAMILY

BUNCHBERRY
Cornus canadensis

Other names: LOW or DWARF CORNEL, PIGEON BERRY, CANADIAN DOG-WOOD

Description: A whorl of leaves, usually 6 in number, fans out from the top of Bunchberry's upright stem. One or 2 pairs of smaller opposite leaves may grow beneath this whorl. The leaves are pointed and are smooth or minutely hairy. The plant is 3 to 9 inches tall. Its blossoms are really 4 petal-like bracts surrounding a cluster of tiny greenish or yellowish flowers. The plant's name comes from the way the *bright-scarlet* berries are bunched together in a cluster on the stem above the leaves. The berries appear in late August.

Habitat: Bunchberries are apt to be found along the edges of evergreen forests rather than in the heart of the forest, where they would not receive enough sunlight. They are also found in swampy areas or waste land. Their range is from Newfoundland to Alaska, south to New Jersey, West Virginia, Colorado, New Mexico and California.

Berry use: The insipid ripe berries are edible when cooked, but great caution should be used in eating them raw.

FLOWERING HERBS • NIGHTSHADE FAMILY

VIRGINIA GROUND CHERRY
Physalis virginiana

Other names: HUSK TOMATO

Description: This is a branched perennial from 1 to 3 feet tall that sprawls and sometimes spreads over quite a large area. It is notable for the husks in which the berries form. The plant's coarse, dull, alternate leaves taper narrowly to the leafstalks. The stems have long, sticky, soft hairs. Five spreading petals, yellow with a brown or purple center, form a bell-shaped flower that hangs from the leaf axil. The round berries form inside a papery, greenish yellow husk, the enlarged calyx, which appears pushed in where it is attached to the stem. The berries are about the size of small cherries and are *red.* They ripen from July to September.

Bunchberry

Habitat: The Virginia Ground Cherry is found in woods and clearings from Manitoba and Ontario to New England, south to Texas and northern Florida.

Remarks: In spite of the name, ground cherries are not related to the cherry family.

Berry use: The unripe fruit and leaves are poisonous. Ripe berries can be eaten raw, but are better cooked into sauce, pie, syrup or jam, with pectin added. The husk with its single berry often drops to the ground before the berry is ripe, but the berry will ripen on the ground. Berries can be gathered and put aside to ripen in their husks. They will then be soft and sweet. Game birds and animals provide competition for the fruits.

Similar species: SMOOTH GROUND CHERRY (*P. subglabrata*) has leaves that are smooth and diamond-shaped, tapering near the leafstalk. It produces berries that are *reddish or purplish.* It is found from Vermont to Washington, south to Florida and Texas. Another species producing red fruit, *P. longifolia,* is native from Nebraska to Texas and west to Arizona. It has been cultivated by the Zuni Indians. The CHINESE LANTERN PLANT (*P. Alkekengi*), a close relative, is cultivated for the sake of its decorative crimson fruit husks. See also Yellow Berries for other ground cherries.

BITTER NIGHTSHADE
Solanum dulcamara

Other names: One of the other names for this vine, BITTERSWEET, causes confusion because it is not related to American Bittersweet, a favorite vine for fall bouquets. Other names for Bitter Nightshade are EUROPEAN BITTERSWEET, CLIMBING NIGHTSHADE, WOODY NIGHTSHADE.

Description: This is a thornless vine, 2 to 8 feet long, with ridged stems and hollow pith. It is an herb except in warm climates, where it becomes a shrub. The leaves are 2 to 4 inches long, with a pointed tip. The shape varies, but usually there are 2 small lobes near the base, 1 on each side. The small, drooping flowers are attractive, with 5 violet (occasionally white) recurved petals; yellow anthers unite to form a cone, or beak. Since they keep blooming late, there are often flowers and

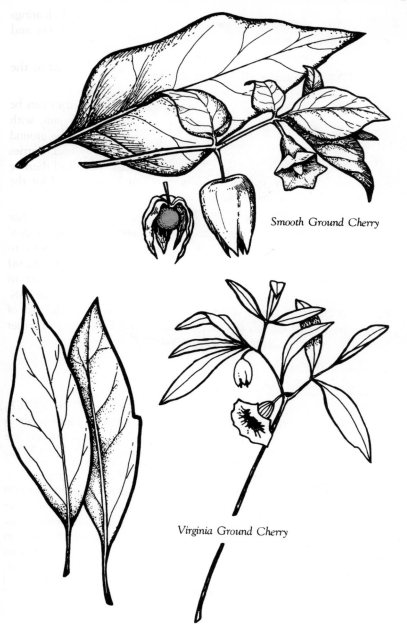

Smooth Ground Cherry

Virginia Ground Cherry

berries on the plant at the same time. The berry color changes from green to yellow to orange to *red*, at maturity. Oval in shape, the berries hang in drooping clusters.

Habitat: This plant originally came from Europe, but it now grows wild nearly everywhere in the United States, especially in waste places, moist thickets and clearings.

Remarks: Except for the opposite-leaved honeysuckles, this is the only twining vine with a hollow pith. It is a relative of the potato, tomato and eggplant, and it sometimes transmits diseases to these plants.

Berry use: The berries are bitter and *poisonous,* and eating them can be fatal. Birds such as ruffed grouse, pheasant, bobwhite and black duck eat the berries and are seemingly immune to the poison, solanine.

Similar species: The berries of the cultivated species, WONDERBERRY (*S. Burbankii*) are reportedly safe to eat if fully ripe and cooked. See also Common Nightshade, in Black Berries, and Purple Nightshade, in White or Green Berries.

FEVERWORT
(*Honeysuckle Family*) See Yellow or Orange Berries.

FLOWERING HERBS • MADDER FAMILY

PARTRIDGEBERRY
Mitchella repens

Other names: TWINBERRY, SQUAWBERRY, TWO-EYED BERRY, RUNNING BOX

Description: This small, creeping, slightly woody herb forms evergreen mats. It has slender, trailing stems that root freely at the nodes. The shining, paired leaves are oval or heart-shaped and blunt at the tip. The white, waxy, fragrant flowers grow in pairs at the ends of the stems, like small funnels spreading out into four lobes. They join together to form a round, *red,* "2-eyed" berry, ⅓ inch or less in diameter, which ripens in the fall. If the berries are not eaten by birds or people, they persist through the winter.

Habitat: Partridgeberry can be found in dry to moist woods and forests

Bitter Nightshade

from Nova Scotia to western Ontario, Minnesota and Arkansas, south to Florida, Texas and California.

Remarks: Since the plants sometimes flower a second time in the autumn, it is possible to see flowers and berries on the same plant. Indian women made a tea of the leaves and drank it before giving birth to ease delivery pains.

Berry use: The berries are dry and seedy, without much flavor, but they are edible and make a colorful addition to salads. Ruffed grouse (often called partridges), bobwhite quails, wild turkeys and such mammals as foxes, raccoons, opossums, squirrels and white-footed mice enjoy the berries.

WOODY VINES AND SHRUBS • YEW FAMILY

AMERICAN YEW
Taxus canadensis

Other names: GROUND HEMLOCK

Description: This is a straggling evergreen shrub about 3 feet tall. It has flat, pointed needles ⅜ to 1 inch long, green on both sides, in flat sprays like those of Balsam Fir. The juicy, berrylike fruit, about ½ inch in diameter, is *red,* with one hard seed exposed at the tip.

Habitat: This shrub grows in moist woods from Canada south to Iowa, Kentucky, West Virginia and New England.

Use of berrylike fruit: Some authorities report that the pulp around the seed is edible. The seeds, however, are *very poisonous.* They contain taxine, a heart-depressing alkaloid. Children have died as a result of eating these berrylike fruits.

COMMON JUNIPER
(Cedar or Cypress Family) See Blue Berries.

WOODY VINES AND SHRUBS • LILY FAMILY

RED-BERRIED GREENBRIER
Smilax walteri

Other names: CORAL GREENBRIER, RED-BERRY GREENBRIER

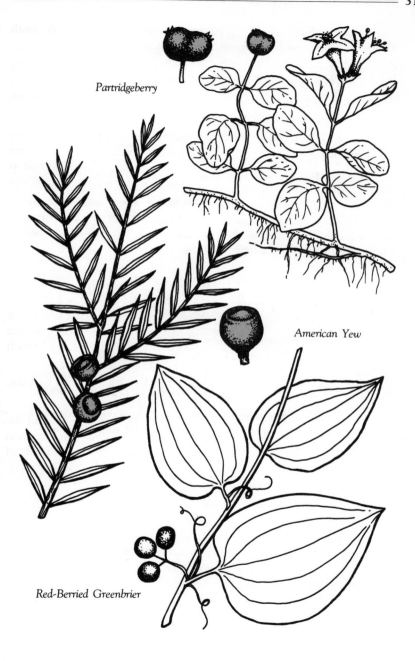

Partridgeberry

American Yew

Red-Berried Greenbrier

Description: See Black Berries for general description of greenbriers. This slender-stemmed, deciduous greenbrier has scattered thorns only toward the base of the plant. The berries are bright *coral red,* about ¼ inch in diameter. They persist through the winter.

Habitat: This is a plant of the coastal plains, from New Jersey south to Florida and west to Louisiana.

Berry use: Although often considered inedible, greenbrier berries can be eaten raw or cooked.

Similar species: LANCELEAF GREENBRIER (*S. smallii*), also called JACKSONBRIER, is a vigorous evergreen vine with a few stout prickles near the base. The leaves are lance-shaped to egg-shaped, dark green and shiny above, paler and duller underneath. The berries are a *dull red* and are edible. Like the Red-Berried Greenbrier, this is a coastal-plains species. See also Black Berries for other greenbriers.

WOODY VINES AND SHRUBS • MISTLETOE FAMILY

MESQUITE MISTLETOE
Phoradendron californicum

Description: This parasitic, shrubby plant has leafless, twiggy stems in large clusters. It is a common parasite on mesquite shrubs. The berries are *red.*

Habitat: This species is found in the Mojave and Colorado deserts.

Remarks: Mistletoe has played an important part in German and Norse mythology. It was believed to bring happiness and good fortune if it did not touch the ground.

Berry use: Mistletoe berries are *poisonous* to humans, but a number of birds and mammals eat them.

Similar species: WESTERN DWARF MISTLETOE (*Arceuthobium campylopodum*) has leaves reduced to small bracts. The berries are *pink;* they explode when touched, shooting the seed a considerable distance. This species is found on conifer branches in the Pacific States. GREENLEAF MISTLETOE (*P. tomentosa*) has thick, leathery leaves on woody twigs that hang from the host tree in masses. Its berries are *pink or white,* with black tips. This species is found in California and Oregon. See also American Mistletoe, White Berries.

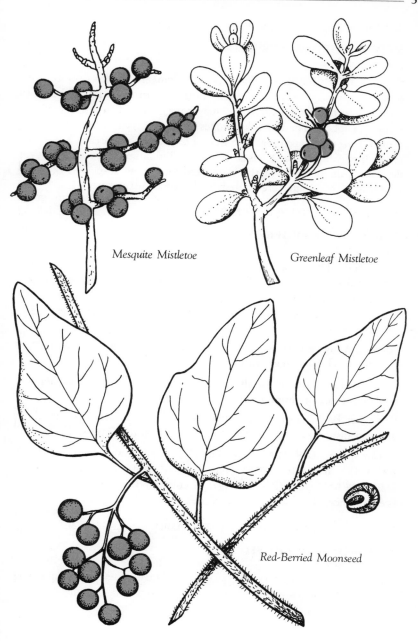

Mesquite Mistletoe

Greenleaf Mistletoe

Red-Berried Moonseed

WOODY VINES AND SHRUBS • MOONSEED FAMILY

RED-BERRIED MOONSEED
Cocculus carolinus

See Common Moonseed, Black Berries, for general description. This scrambling vine is evergreen in the southern part of its range. Its alternate, untoothed leaves vary: they may be egg-shaped, triangular, heart-shaped or lobed. The berrylike drupes are round, pea-sized and *red* with a rigid, crescent-shaped seed like that of Common Moonseed. This shrub grows from Virginia to Kansas, south to Florida and Texas. *Do not eat this fruit!* All members of the moonseed family are suspected of being *poisonous.* The single, crescent-shaped seed is distinctive.

WOODY VINES AND SHRUBS • BARBERRY FAMILY

AMERICAN BARBERRY
Berberis canadensis

Description: This is a deciduous shrub 3 to 6 feet tall, armed with 3-pronged spines at the leaf axils. The leaves are egg-shaped, tapering to a point near the base, and sharply toothed. The yellow flowers bloom in a cluster and are replaced in September or October by *bright red or scarlet* berries.

Habitat: American Barberry is found in Missouri and in mountain woods from Pennsylvania south to Georgia.

Remarks: This shrub is an alternate host of wheat rust and is therefore regarded as an undesirable plant.

Berry use: Barberries make excellent jelly; they are rich in pectin. They are also tasty served as cooked fruit or made into a cold drink.

Similar species: EUROPEAN BARBERRY (*B. vulgaris*), also called COMMON BARBERRY and SOURBERRY, has escaped cultivation in the East. JAPANESE BARBERRY (*B. thunbergii*) is cultivated as a hedge because of the ornamental value of its *red* berries.

WOODY VINES AND SHRUBS • LAUREL FAMILY

COMMON SPICEBUSH
Lindera benzoin

Other names: ALLSPICE BUSH, WILD ALLSPICE, BENJAMIN BUSH, FEVERBUSH, SPICEBUSH

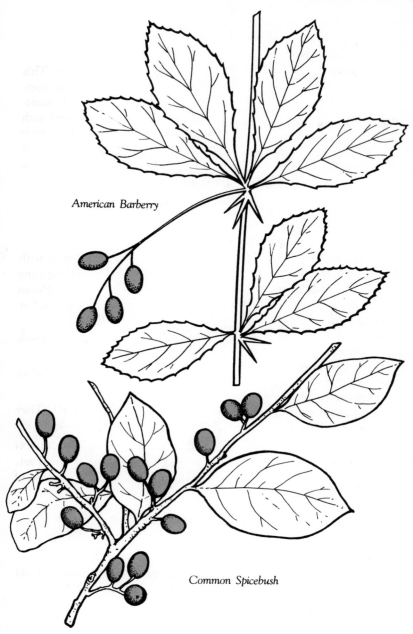

American Barberry

Common Spicebush

Description: This deciduous shrub, which grows to 15 feet tall, is notable for the spicy aroma of all its parts. Even in midwinter the spicy scent can be obtained by breaking a twig. The leaves are alternate, elliptic and untoothed. The small, yellow, fragrant flowers appear before the leaves do, and are clustered close to the branch. The oval berries are *bright red* and remain on the bush after the leaves have fallen.

Habitat: Common Spicebush grows in damp woods and stream banks from New England and Ontario south to Florida and Texas.

Berry use: The dried, powdered berries can substitute for allspice.

WOODY VINES AND SHRUBS • SAXIFRAGE FAMILY

CURRANTS
Ribes spp.

Currants are deciduous shrubs with small, alternate, maplelike leaves clustered on short "spur" branches. Small, bell-shaped flowers are followed by pulpy, round berries tipped with the remains of the calyx. Together with gooseberries, to which they are closely related, currants are alternate hosts to blister rust, which affects white pine trees. For this reason, they have been eradicated in and near areas of our coniferous forests and are not now as common as they once were. They can be distinguished from gooseberries by the fact that they usually lack thorns and the berries grow in long clusters, separate easily from their stalks and are generally smooth-skinned.

There are numerous kinds of currants; one species or another grows nearly everywhere in North America. The berries are never poisonous, but some are much better-tasting than others. They are all rich in pectin, especially when slightly underripe. Some species with red berries are as follows:

GARDEN RED CURRANT
Ribes sativum

Description: Reddish shredded bark on the stems is a characteristic of this 3-foot shrub. The leaves are 3-lobed with toothed edges. The flowers are yellow-green, the berries *red*, translucent, smooth and juicy.

Habitat: This shrub has escaped from cultivation to wasteland throughout northern United States and southern Canada.

Garden Red Currants

Swamp Red Currants

Berry use: Although the berries can be eaten raw, they are best cooked into sauce, jelly or pie fillings.

Similar species: SKUNK CURRANT (*R. glandulosum*) is an eastern species easy to recognize because all parts of it give off a skunklike odor when crushed. The bristly *coral-red* berries have an unpleasant taste. SWAMP RED CURRANT (*R. triste*), also an eastern species, has brownish or purplish flowers that hang in drooping clusters. The berries are smooth and *bright red.* ALPINE PRICKLY CURRANT (*R. montigenum*) and WAX CURRANT (*R. cereum*), also called SQUAW CURRANT, are found in the mountainous regions of the West. See also Purple Berries, Yellow or Orange Berries and Black Berries for other currants.

BRISTLY GOOSEBERRY
Ribes setosum

See Purple Berries for general description and illustrations of gooseberries. This species, also called ROCK GOOSEBERRY, has reddish brown bristly branchlets, sometimes armed with thorns. The edible berries are *red or black*, smooth or slightly bristly. This species grows from Ontario to Alberta, south to Michigan, Wisconsin and Nebraska.

WOODY VINES AND SHRUBS • ROSE FAMILY

RED CHOKEBERRY
Pyrus arbutifolia

See Black Chokeberry for general description. Woolly branchlets and leaf undersides characterize this shrub. The ¼ inch *red* berrylike pomes persist well into winter. Red chokeberry grows in damp areas from Nova Scotia to Ontario, south to Missouri, Texas and Florida. These tasty fruits can be used like blueberries.

WOODY VINES AND SHRUBS • ROSE FAMILY

RASPBERRIES AND BLACKBERRIES
Rubus spp.

These shrubs are compound-leaved, soft-wooded, short-lived, and are usually bristly or prickly. Their simple stems are called canes, and new ones grow each year, which is good since a cane produces flowers and fruit the second year and then dies. The leaves are long-stalked, alternate, divided into 3 or more leaflets. The flowers are 5-petaled and

Red Chokeberry

usually showy. Although we call them berries, the fruits are actually dense clusters of 1-seeded drupelets. All are edible.

Raspberries are easily separated from their receptacles, and after being picked, they have a hollow inside. This distinguishes them from blackberries, which are firmly attached to their receptacles. Raspberries, blackberries and dewberries are all called brambles, which is regarded as the most valuable wild fruit crop in America. Over 150 kinds of birds and mammals, including big game animals, eat the twigs and berries of brambles. These plants are also valuable in soil-erosion control, and they provide cover for wildlife.

RED RASPBERRY
Rubus idaeus

Description: This arching shrub has round, bristly stems, sometimes slightly whitened. The 3 to 5 leaflets are egg-shaped, double-toothed, whitened and downy underneath. White flowers appear from May to July. The *red* fruits, which need no description, ripen from July to September.

Habitat: Wild raspberries grow in waste areas from Newfoundland to British Columbia, south to North Carolina and California.

Use of berrylike fruit: Raspberries can be cooked into sauce, jam or pie, but they are so delicious raw that many people prefer them this way.

BLACK RASPBERRY, DEWBERRY, BLACKBERRY
See Black Berries.

PURPLE-FLOWERING RASPBERRY
Rubus odoratus

Description: This is a rambling, thornless shrub about 3 to 6 inches tall, with large, maplelike leaves. The purplish flowers are showy, 1 to 2 inches broad. *Dull-red* fruit, shaped like a flat raspberry, ripens from July to September.

Habitat: Look for this plant in rocky woods and thickets from Nova Scotia to Michigan, south to Georgia and Tennessee.

Use of berrylike fruit: The edible fruits are sour and dry, improved by cooking.

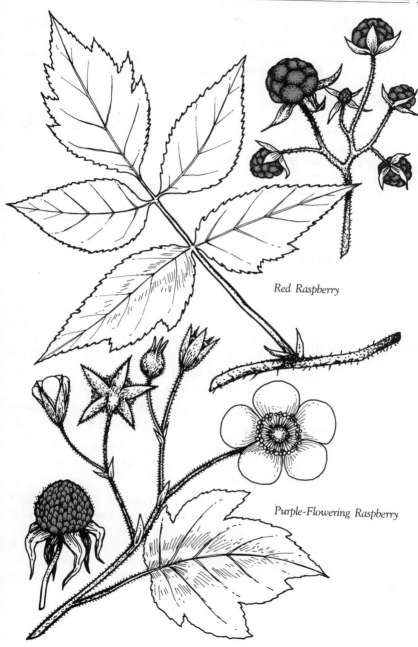

Red Raspberry

Purple-Flowering Raspberry

Similar species: THIMBLEBERRY (*R. parviflorus*), also called SALMON BERRY or WHITE-FLOWERING RASPBERRY, is shorter, has white flowers and is rarely bristly. The *orange-red* berries have a bland flavor but are juicy and are often used to make jam or jelly.

WILD ROSES
Rosa spp.

Like raspberries and blackberries, roses are arching shrubs that are usually prickly or bristly. They have 3 to 11 sharply toothed leaflets per leaf. At the base of each leafstalk are "wings" (stipules). Twigs and stems are usually green or red. The 5-petaled, showy, fragrant flowers have been the subject of much praise in song and poetry. The fruits, known as hips, are *usually red* and fleshy, with numerous small seeds, and they are edible. There are many species of wild rose, distributed in most parts of the United States.

PASTURE ROSE
Rosa carolina

Other names: WILD ROSE, CAROLINA ROSE, LOW ROSE

Description: Straight, slender thorns protect this erect shrub, which grows ½ to 2½ feet tall. The leaves usually have 5 elliptic or egg-shaped leaflets, sharply toothed, dull above, slightly downy underneath. Solitary pink flowers, about 2 inches across, bloom from May to July. The fruit is *red*, roundish, puckered and tufted at the top, and covered with glandular hairs. It clings to the bush throughout the winter.

Habitat: This species grows in dry, open woods, thickets and pastures from Nova Scotia to Minnesota, south to Florida and Texas.

Use of berrylike fruit: Rose hips, which are rich in Vitamin C, are often steeped in hot water to make tea. The hips are also used for jelly (with pectin or apples added), candy and emergency food.

Similar species: NOOTKA ROSE (*R. nutkana*) grows in evergreen forests from the Rocky Mountains to Alaska and south to California. SWEETBRIAR (*R. eglanteria*), also called EGLANTINE, grows to 10 feet tall, and is found throughout the United States. It has 7 to 9 leaflets in the compound leaf and curved prickles on the stem at the base of each leaf. The brilliant *red* fruits appear in late summer. They can be used like other rose hips.

Pasture Rose

CHERRIES AND PLUMS
Prunus spp.

Most wild cherries are shrubs or small trees. Their fruits usually look like berries, though they are really drupes, with a bony pit enclosing the seed. Shrubs or trees in this genus have alternate, simple leaves, single-toothed and narrow at the base. Broken twigs have an odor resembling almond; this is one good clue to identification. Cherries are thornless and their fruits do not have a white powder on them. Many of the plum species (sometimes called Indian cherries) have thorns, and the fruits have a white, powdery bloom. Fruits of all species of cherry and plum are eaten by many kinds of birds and animals. Man finds them all edible, though the seeds of cherries are poisonous. One species or another of the prunus genus can be found through the United States. Here are some of the cherry and plum trees with red fruits:

CHOKECHERRY
Prunus virginiana

Other names: COMMON CHOKECHERRY, WILD CHERRY

Description: Frequently a large shrub, Chokecherry can also become a small tree, up to 25 feet tall, with a trunk 8 inches in diameter. The bark is fissured into small scales and has an acrid odor when crushed. The leaves are egg-shaped, with fine, sharp teeth. Long, dense clusters of large, white, strongly scented flowers bloom from April to June. The pea-sized fruits are round, *dark red to purplish black,* bending the branches beneath their weight. They ripen from July to September.

Habitat: This may be the most widely distributed shrub or tree in North America. It grows in open woods, fields, thickets, and along fences and streams. It is found everywhere in the United States except the southeastern and south central states.

Remarks: A tea made from the bark of the Chokecherry tree was long used in Appalachia to treat measles and colds. Indians used a similar tea to ease childbirth pains.

Use of berrylike fruit: Chokecherries are too sour and puckery to be very desirable raw. If they *are* eaten raw, the *pits must not be swallowed,* since they are poisonous. Both wilted leaves and *fresh seeds of all wild cherries contain cyanide.* However, cooking destroys the cyanide, so the pits do

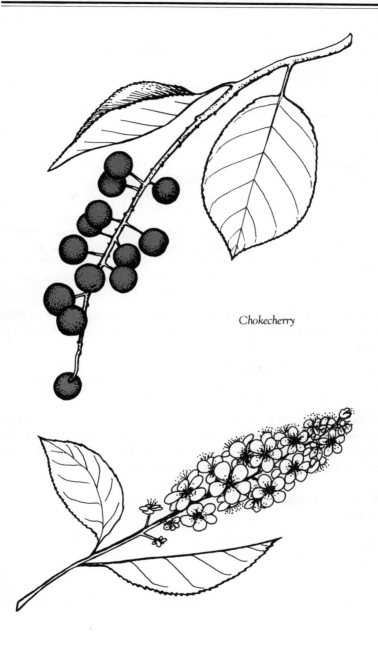

Chokecherry

not need to be removed before boiling the berries for jelly. With the addition of pectin, wild cherries make delicious jelly.

PIN CHERRY
Prunus pensylvanica

Other names: FIRE CHERRY, WILD RED CHERRY, BIRD CHERRY

Description: This is an early-blooming shrub or small tree with a short trunk, narrow crown and slender horizontal branches. Its narrow leaves have in-curved teeth and are bright green and shiny above, paler beneath. The white flowers, in umbrellalike clusters, appear when the leaves are half-grown. Pea-sized fruits ripen from July to September. They are *light red*, thick-skinned, with thin flesh.

Habitat: This fast-growing, short-lived tree often takes over completely in recently burned areas — hence the name Fire Cherry. It also grows in thickets and young woods from Labrador to British Columbia, south through New England to Virginia and the mountains of Georgia and Tennessee, and westward to include most of the Midwest.

Remarks: Chokecherry twigs simmered in water were used to successfully treat a digestive disturbance of Captain Lewis during the Lewis and Clark Expedition of 1804— 6.

Use of berrylike fruit: These cherries are too sour to eat fresh, but their color and flavor are such that jelly made from them is delicious. See Chokecherry remarks, just preceding.

Similar species: BITTER CHERRY (*P. emarginata*) is a western species that greatly resembles Pin Cherry in its flowers and fruit, but its leaves are wider and not so pointed. It grows from British Columbia to Montana and south to California, Arizona and New Mexico.

BLACK CHERRY
See Black Berries.

AMERICAN PLUM
Prunus americana

Other names: WILD PLUM

Description: This shrub or small tree often has spine-tipped branches.

Pin Cherry

The bark of mature trees is scaly. The oval leaves have a tapering tip and toothed margins and are smooth, or nearly so, on both surfaces. White flowers bloom in a cluster of 2 to 5. Round, *red or yellow* fruits, ¾ to 1 inch in diameter, ripen from August to October.

Habitat: This species often forms thickets along fence rows, streams and wood borders. Its range is roughly the eastern two-thirds of the United States.

Use of berrylike fruit: Though tart, these plums can be eaten raw. They make excellent sauce or jam. Some plums do not require pectin for jelly; most do.

Similar species: CANADA PLUM *(P. nigra)* and CHICKSAW PLUM *(P. angustifolia)* are similar to American Plum. Canada Plum grows in the Northeast and the Midwest, Chicksaw Plum in southeast and south central states. HORTULAN *(P. hortulana)* and WILDGOOSE PLUM *(P. munsoniana)*, eastern species, are also similar. KLAMATH PLUM *(P. subcordata)*, also called WESTERN PLUM, grows in Washington and Oregon. All of these species have *red or yellow* fruits.

CHRISTMAS BERRY
Photinia arbutifolia

Other names: TOYON BERRY

Description: A shrub or small tree, this evergreen grows to 30 feet tall. Its leaves are dark green, shiny, leathery and toothed. The white flowers are small and numerous. The berrylike fruit is *red, sometimes yellow,* and pulpy.

Habitat: This shrub grows on low mountain slopes in California.

Use of berrylike fruit: The fruits are edible raw or cooked, but are best with sugar added.

WOODY VINES AND SHRUBS • CROWBERRY FAMILY

PURPLE CROWBERRY
Empetrum atropurpureum

This is similar to Black Crowberry except that the trailing branchlets and leaves are white and woolly when young. The berries are *red to purplish black.* See Black Crowberry in Black Berries.

American Plum

Chicksaw Plum

Christmas Berry

WOODY VINES AND SHRUBS • CACTUS FAMILY

PRICKLY PEAR
Opuntia spp.

Other names: INDIAN FIG, DEVIL'S TONGUE

Description: Prickly Pears have fleshy rather than woody stems, but they are persistent over the winter, so they are usually classified as shrubs. There are two main groups of the opuntia genus: those with flat, jointed pads (the Prickly Pears), and those with cylindrical, jointed stems (the cholla, or cane, cacti).

The jointed, pear-shaped pads of which the Prickly Pear is composed form its stem. Its scalelike leaves quickly fall off. Small, brown, barbed hairs are evenly spaced over the pads. These hairs are easily detached from the plant, but because of the barbs, they may become painfully imbedded in the skin. The 1 or more sharp spines, up to ½ inch in length, look more dangerous but are actually less so. Short-stemmed flowers grow at the edges of the pads. They are yellow or reddish, large and showy, bringing great beauty from May to July in the dry countryside where they may be seen. The fruit is a berry, an inch or more in length, usually *red or purple* when mature. It is called tuna by Southwesterners.

Habitat: Sandy soil, rocky places, deserts and coastal shrub areas provide the right environment for Prickly Pears. One species or another grows from Massachusetts to Minnesota, south to Florida, Texas, Arizona and California.

Remarks: Cacti were originally confined to the Americas. Although we often think of cacti as desert plants, they are native to all of the United States except Maine, New Hampshire, Vermont, Alaska and Hawaii. Several species of opuntia are endangered or threatened.

Berry use: Most Prickly Pears are edible. The pulp of the berry between the skin and its seeds is tasty, especially when chilled, but the barbed hairs must first be removed. This can be done with a damp cloth. Gloves should be worn when handling the fruit. The dried seeds can be ground for use as a flour or soup thickener. The berries of eastern species are seedy and insipid compared with those that grow in the southwestern desert country. Many animals also eat Prickly Pears.

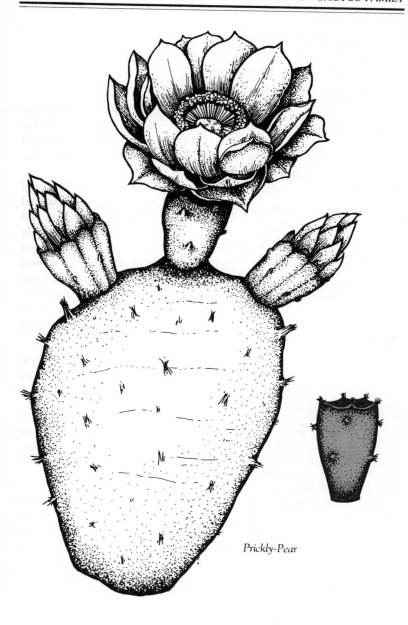

Prickly-Pear

DESERT CHRISTMAS CACTUS
Opuntia leptocaulis

Other names: HOLYCROSS CHOLLA, CHRISTMAS CHOLLA, DARNING-NEEDLE CACTUS, PENCIL-POINT CACTUS.

Description: This cactus belongs to the cholla (cane) group of opuntia. It is a slender-stemmed, spiny plant whose many branches become intertangled to form a bush 2 to 4 feet tall. Greenish or yellow flowers are so sparse and inconspicuous that they may be overlooked. The berries, however, are very noticeable, for they are ½ to 1 inch long, round and scarlet. They remain on the stems through most of the winter.

Habitat: Desert areas in Arizona, Oklahoma and south to Mexico are home to this cactus.

Remarks: The manufacture of food in this cactus, as in other cacti, is primarily carried out by chlorophyll in the green cells of the stems.

Berry use: The berries of this cactus are not reported as being edible, though the berries of some chollas are.

PINCUSHION CACTUS
Mammillaria spp.

Other names: NIPPLE CACTUS, BUTTON CACTUS, BABY CACTUS, FISHHOOK CACTUS

Description: These cacti have ball-like stems that are from 1 to 12 inches in height and diameter and are persistent over the winter. The scientific name comes from the Latin word for nipple and refers to the nipplelike knobs (tubercles) of the plant. Each of these tubercles is crowned with a cluster of slender, needlelike spines. Usually one or more of the central spines in each cluster is longer than the others and is hooked. The relatively small flowers, white, yellow, red or purple, are arranged in an attractive geometric pattern. The berries are naked and smooth, club-shaped or round. They may be *red, green, yellow or dull purple,* and are borne at the base of the tubercles.

Habitat: Members of this genus grow in desert or arid grasslands from southeast California to Texas.

Berry use: Some of the members of this genus, SLENDER PINCUSHION CACTUS (M. *fasciculata*) for instance, have edible berries.

Desert Christmas Cactus

Pincushion Cactus

Similar cacti: HEDGEHOG CACTUS (*Echinocereus spp.*), also called CLARET CUP CACTUS, KING'S CUP, STRAWBERRY CACTUS, is a small ball-shaped, spiny cactus with red berries that are edible once the spines are removed.

WOODY VINES AND SHRUBS • MEZEREUM FAMILY

LEATHERWOOD
Dirca palustris

Other names: MOOSEWOOD

Description: The bark of this shrub is so tough that it cannot be broken by hand. The shrub, growing to 8 feet in height, often has only one trunklike stem. The branchlets, tough and pliable, appear jointed. Leaves are alternate, short-stalked, oval to elliptic. The small yellow flowers appear before the leaves. Ripening early, in May or June, the berrylike fruits are fleshy and red.

Habitat: Leatherwood grows where the soil is sandy or rocky, from Newfoundland to Alaska, south to New England, the Great Lakes region and South Dakota.

Remarks: The tough bark of this tree was used by Indians for bowstrings, fishlines and baskets. The roots and bark were used medicinally.

Use of berrylike fruit: The fruit does not appear in the lists of either edible or poisonous fruits. Caution is advised. The bark contains poisonous substances that can cause irritation and blistering of the skin of those susceptible.

DAPHNE
Daphne mezereum

Description: This is a low shrub, 3 feet or less in height, with very tough bark. Its leaves are wedge-shaped, alternate and untoothed. Purplish flowers in clusters appear before the leaves. The berrylike fruits, scarlet, leathery and single-seeded, ripen in August or September.

Habitat: This introduced shrub has escaped cultivation, especially on limestone soils, from Newfoundland and Ontario to New York, west to Ohio.

Leatherwood

Daphne

Remarks: This and similar species are widely cultivated as ornamentals.

Use of berrylike fruits: Poisonous! All parts of this plant, but particularly the attractive berrylike fruits, contain a glycoside that causes ulceration of the throat and stomach, vomiting, internal bleeding, coma and death. Even a few fruits can be fatal to a child. Birds seem to be immune to the poison.

WOODY VINES AND SHRUBS • CASHEW FAMILY

STAGHORN SUMAC
Rhus typhina

Description: Very hairy twigs and leafstalks distinguish this shrub, which sometimes grows to a small tree. The leaves are composed of 11 to 31 lance-shaped, toothed leaflets. They turn a lovely scarlet very early in the fall. The twigs are stout and give forth a milky sap when cut. Greenish male and female flowers grow on separate shrubs in dense, erect clusters at the ends of the branches. The berrylike fruits are *red,* hairy drupes grouped together in long, erect clusters at the branch ends. They mature in August or September and remain through the winter, giving the shrub an unmistakable appearance.

Habitat: Staghorn Sumac grows in dry soil along fences, in waste places and open lots from Nova Scotia to Minnesota, south to Georgia, Illinois and Iowa.

Remarks: The young branches, covered with brown hairs, have an appearance similar to the fuzzy new horns produced by stags in spring, resulting in the name "staghorn."

Use of berrylike fruit: The Indians soaked the berries in water and sweetened the resulting liquid with maple syrup to make a beverage that tasted much like lemonade. Colonial housewives used the unsweetened liquid as a substitute for lemon juice or vinegar. From personal experience in making "Indian lemonade," I would recommend frequent sampling while the fruit is steeping. Properly prepared, the beverage is as tasty as real lemonade, but if the fruit is allowed to steep too long, the taste becomes bitter. The berries can be dried for use in winter. White-tailed deer and numerous birds eat sumac fruits.

Staghorn Sumac

Smooth Sumac

Similar species: SMOOTH SUMAC *(R. glabra)*, also called SCARLET SUMAC or UPLAND SUMAC, has hairless twigs and leafstalks and is smaller, growing to only 15 feet. Western species include LEMONADE SUMAC *(R. integrifolia)*, with hairy leaves divided into 3 leaflets, SUGAR SUMAC *(R. ovata)* and KEARNEY SUMAC *(R. kearneyi)*, both with evergreen leaves. All have red berrylike fruits. See also White or Green Berries for Poison Sumac.

WOODY VINES AND SHRUBS • HOLLY FAMILY

HOLLIES
Ilex spp.

The holly family contains a very diverse group of shrubs and trees. Some are evergreen, some are leaf-losing. The leaves are simple and alternate. The small, greenish white flowers are followed by round, berrylike fruits with 4 to 6 bony nutlets and mealy or pulpy flesh. *Holly berries are poisonous to humans.* Here are some holly shrubs with red berrylike fruits:

COMMON WINTERBERRY HOLLY
Ilex verticillata

Other names: WINTERBERRY, VIRGINIA WINTERBERRY, BLACK-ALDER

Description: This is a deciduous shrub 3 to 15 feet tall, with smooth, grayish bark. The dark-green leaves are lance-shaped or nearly circular, slightly leathery and coarsely toothed. The berrylike fruits are shiny and fiery scarlet *(occasionally yellow)*. They are short-stemmed, usually in pairs. No other member of the holly family has berries so brilliant or in such abundance. They ripen in September and October and persist after the leaves fall.

Habitat: Exclusively American, this shrub grows along streams and in swamps and wet woodlands from Newfoundland to Minnesota, south to Georgia and Missouri.

Use of berrylike fruits: The fruits are somewhat *poisonous* and can cause vomiting, diarrhea and stupor if eaten in quantity. They are especially dangerous for small children. However, more than 48 species of birds

Winterberry Holly

Dahoon Holly

Mountain-Holly

and various small mammals eat them. Some observers say the birds will eat Winterberries only when other food is unavailable.

Similar species: DECIDUOUS HOLLY (*I. decidua*), also called POSSUM-HAW, is a shrub or tree with *red* berrylike fruit. DAHOON HOLLY (*I. cassine*) is an evergreen shrub or tree that grows to 25 feet. Its small fruits are *red, occasionally yellow.* YAUPON HOLLY (*I. vomitoria*), also called CHRISTMAS-BERRY, is much like Dahoon Holly. Its caffeine-containing leaves were once used by Indians to make a purgative ceremonial "black drink." MOUNTAIN HOLLY (*Nemopanthus mucronatus*) is related to the hollies but differs from them in that it lacks the appendages at the base of the leafstalk called stipules. Its berries are *dull red.* See also American Holly, Trees, Red Berries.

BURNING-BUSH
Euonymus atropurpureus

Other names: WAHOO, EASTERN WAHOO

Description: Burning-Bush is a shrub or small tree with green, 4-lined twigs. The leaves are egg-shaped or elliptic, short-stalked and fine-toothed. Purple flowers hang in clusters from the leaf axils. The fruits are really 4-lobed capsules with seeds, but they are berrylike and attractive. When the *pinkish purple capsule* splits open, it reveals the *orangish red seed covering* (aril). The effect resembles that of American Bittersweet.

Habitat: This shrub grows in moist woods, along streams and roadsides from New York and southern Ontario to Oklahoma, Alabama and Virginia.

Use of berrylike fruit: The fruit can be used in fall bouquets as you would use American Bittersweet. All members of the Bittersweet family are suspected of being *poisonous.* Children are definitely poisoned by Burning-Bush seeds if they are eaten in quantity.

Similar species: STRAWBERRY-BUSH (*E. americana*), also known as BURSTING-HEART, is more straggling and the branchlets are 4-sided. Its *crimson* fruits have been reported as being *poisonous* to children, and even most birds do not seem to relish them.

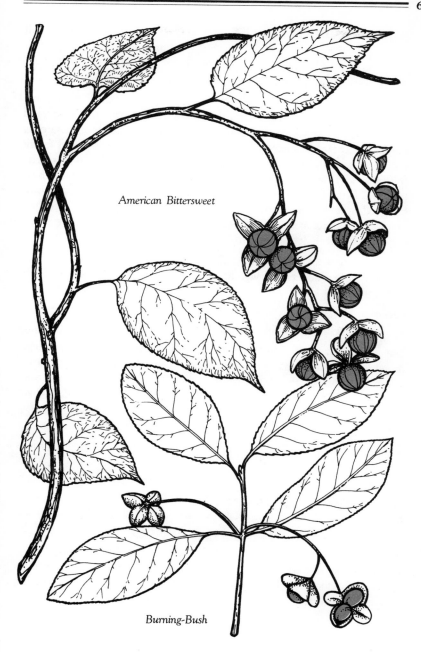

American Bittersweet

Burning-Bush

AMERICAN BITTERSWEET
Celastrus scandens

Other names: CLIMBING BITTERSWEET, WAXWORK, SHRUBBY BIT-
TERSWEET. Deadly Nightshade is also called Bittersweet, but it should
not be confused with American Bittersweet.

Description: This is a high-climbing, tangled woody vine that may
scramble over low vegetation, trail on the ground or climb on trees to a
height of 60 feet. Its leaves are oval or egg-shaped, pointed and finely
toothed, dark green above and paler beneath. Its stems are round,
thornless and hairless, and its buds are unusual in that they are set at
nearly right angles to the stems. The small, yellowish green flowers are
followed by *dull-orange* capsules, about ⅜ inch in diameter. When they
mature in September or October, they expose the *bright scarlet* seed
covering (aril). If left alone, the fruits persist through the winter.

Habitat: American Bittersweet grows along roadsides, fence rows, river-
banks, and in thickets and woodlands from Quebec to Manitoba, south
to Georgia and Oklahoma.

Remarks: Horses, sheep and cattle are occasionally poisoned by eating
the leaves, which contain a toxic substance.

Use of berrylike fruit: The fruits are *poisonous* for humans. Songbirds,
ruffed grouse, pheasants, quail, rabbits and squirrels eat them without ill
effects. Because of its colorful fruits, American Bittersweet is a favorite
vine for use in fall bouquets. In many areas it has been overcollected to
such an extent that it has been greatly reduced or even exterminated.

HOLLYLEAF BUCKTHORN
Rhamnus crocea

This member of the buckthorn family is an evergreen shrub or small
tree with *red* berries. They have a laxative effect, and it is reported they
will turn the skin red temporarily if eaten in quantity. This species
grows in California and Arizona. See Black Berries for a general discus-
sion of buckthorns.

Hollyleaf Buckthorn

Devil's-Club

WILD GRAPE
(Vine Family) See Purple Berries.

SILVERY BUFFALOBERRY
Shepherdia argentea

Similar to Canada Buffaloberry, this species grows farther west, in the western third of the United States. It is sometimes thorny and has leaves that are silvery on both sides. The berrylike fruits are *bright red or golden* and have a pleasantly tart flavor. They make fine jelly and may be dried for future use or eaten raw. See Canada Buffaloberry, Yellow or Orange Berries.

DEVIL'S-CLUB
Oplopanax horridus

This shrub is related to Hercules'-Club and is similar to it except that its berrylike fruit is *red.* The plant and its berries are considered *poisonous.* It ranges from Ontario and Isle Royale in Michigan to Alaska, south to Montana and California. See Hercules'-Club, Black Berries.

WINTERGREEN
Gaultheria procumbens

Other names: CHECKERBERRY, TEABERRY, CREEPING or RED-BERRY WINTERGREEN, BOXBERRY

Description: This entire plant has a strong odor of wintergreen. It is a creeping, evergreen, semiherbaceous shrub with erect, branchlike stalks that arise from trailing or underground stems, often spreading to an area of considerable size. New shoots in the spring are red. The thick, shiny leaves are oval and slightly toothed. They turn bronze in winter. Flowers form near the ends of the branches, solitary or in clusters, dangling beneath the leaves. The white petals fuse to form an urn-shaped, small

Wintergreen

flower. Small *red* berries appear in August and may remain on the plant until next June.

Habitat: Look for Wintergreen plants in woods and open places, especially under evergreen trees. They are most abundant in sandy or rocky soil, and are found from Newfoundland to Manitoba and southward.

Remarks: Because partridges and ruffed grouse are fond of the berries, this plant is known in some areas as Partridgeberry, but this name is more properly given to *Mitchella repens*. (See Partridgeberry, p. 28) The Indians and explorers made a tea from the leaves of Wintergreen and used it as a remedy for rheumatism and other ailments. Wintergreen flavor is now obtained from the bark of black birch or produced synthetically.

Berry use: The berries have a pleasantly spicy, refreshing flavor and can be eaten raw or used in salads or cooking. To obtain maximum flavor when the berries are used in cooking, they should be crushed and added to the cooked ingredients. In addition to grouse and partridges, wild turkeys, deer and bears also enjoy Wintergreen berries.

Similar species: There are several species of Wintergreen, including some western varieties. See Purple Berries for a species known as Salal.

BEARBERRY
Arctostaphylos uva-ursi

Other names: KINNIKINNICK, EVERGREEN BEARBERRY, MANZANITA, MEALYBERRY, BEAR'S-GRAPE, HOG BERRY, BEAR MAT, HOG CRANBERRY

Description: This is an attractive low, trailing evergreen shrub that forms a ground cover about 6 inches high. Its long, flexible branches may root at the nodes. They are smooth, woolly or sticky-downy. The short-stalked, alternate leaves are paddle-shaped, smooth and leathery, glossy above. Small, waxy, white or pinkish flowers, shaped like narrow-mouthed urns, bloom in terminal clusters. Ripening in August or September, the berries may persist through the winter. They are round, *red*, about ¼ inch in diameter. The flesh is mealy and contains 5 to 10 seeds.

Habitat: Sandy open spots, tundras or rocky areas provide the right en-

Bearberry

vironment for Bearberries. They grow above the treeline in the mountains. Their range is from the Canadian Arctic south to Virginia, New Mexico and California.

Remarks: Indians mixed the dried leaves with tobacco, calling the resulting product Kinnikinnick, and smoked it in their pipes. (They also used other plants for this purpose.) The leaves have also been used medicinally, for tanning leather and for making dyes.

Berry use: Although they are edible and nutritious, with a high vitamin C content, the berries have tough skins, are dry and not very appetizing raw. Cooking improves the flavor. They are often combined with other berries for sauce or jelly. Bearberries are best picked after the first frost. Bears emerging from hibernation eat the berries and leaves. Grouse and wild turkeys are fond of the berries.

Similar species: RED BEARBERRY (A. *rubra*) has leaves that are much wrinkled. The berries are *scarlet* and juicy. This species is found from Alaska to Newfoundland and in southern Canada. For Alpine Bearberry, see Purple Berries.

MANZANITA
Arctostaphylos spp.

Description: Manzanitas are shrubs with tough, crooked branches and polished, dark-red bark, which peels off in thin curls. They grow several feet tall, erect or spreading, and usually have fuzzy twigs. The evergreen leaves are oval and pointed. The flowers clustered at the branch ends vary in color — creamy, pinkish white or yellowish brown. The berries are *red or brown*, sometimes fuzzy and mealy.

Manzanitas are western shrubs related to Bearberries. They include PARRY MANZANITA (A. *manzanita*), GREENLEAF MANZANITA (A. *patula*), POINTLEAF MANZANITA (A. *pungens*), WHITELEAF MANZANITA (A. *viscida*) and PINEMAT MANZANITA (A. *nevadensis*).

Habitat: Growing from moderate altitudes to 8,000 feet, Manzanitas are found on dry, but not desert, ground throughout the West. They reach their peak in California, where about 25 species are recognized.

Berry use: The berries are usually regarded as survival food. Indians used them in making pemmican and cider. The berries of Whiteleaf Man-

Greenleaf Manzanita

zanita are sticky, hard to handle, and do not have much flavor. Those of Greenleaf Manzanita are dry, seedy and very sour, but they can be made into a refreshing drink by boiling, straining and sweetening them. Harvesting is often difficult because of the dense thickets formed by this shrub. The berries of Pinemat Manzanita, a sprawling shrub that spreads out like a ground cover, may be used to make cider without the addition of sugar.

GROUSE WHORTLEBERRY
Vaccinium scoparium

This is a western species that resembles a typical Blueberry except that the berries are *deep red* and look as if the ends had been cut off. See Blueberries in Blue Berries.

CRANBERRIES
Vaccinium spp.

Cranberries belong to the same genus as blueberries. They are trailing evergreen shrubs with smooth, hairless leaves and scaleless stems. The flowers are small, white or pink, bell-shaped. The fruits are round, translucent, *red* berries with many seeds. Early colonists called these plants Craneberries because the shape of the blossom and the way it droops at the end of its stem reminded them of the head, beak and neck of a crane. The name has been shortened through the years. Cranberries were a very important food for the colonists because they are so firm that they keep through the winter. They are still an important food in this country.

NORTHERN MOUNTAIN-CRANBERRY
Vaccinium vitis-idaea

Other names: LINGONBERRY, MOUNTAIN CRANBERRY, BOG CRANBERRY, ROCK CRANBERRY, RED BILBERRY, COWBERRY, FOXBERRY

Description: This is a mat-forming evergreen shrub with underground stems and upright branches 1 to 6 inches high. The leaves are small, leathery, oval, glossy on top, pale with small black dots underneath. The white or pink bell-shaped blossoms have 4 short lobes and grow in terminal clusters. The pulpy berries, *dark red*, ½ to ⅜ inch in diameter,

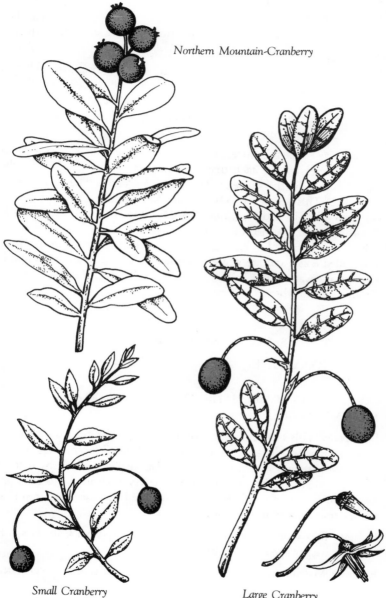

Northern Mountain-Cranberry

Small Cranberry

Large Cranberry

ripen in August or September. Their flavor differs somewhat from that of other cranberries.

Habitat: This species grows from the Arctic region south to New England, Minnesota and British Columbia.

Remarks: Cranberry sauce served at Thanksgiving and Christmas feasts is traditional in this country. Cranberries are also used in relishes, salads, drinks, jellies, pies, and as an addition to hot breads. Wild cranberries are best after the first frost. They keep well under a snow cover and can be used as an emergency food.

Similar species: LARGE CRANBERRY (*V. macrocarpon*) differs from the preceding species by the fact that its petals curve back, exposing the "beak" formed by the stamens. The leaves are blunt. The berries are a *lighter red.* This species is found from Newfoundland to Minnesota, south to Illinois and Virginia. SMALL CRANBERRY (*V. oxycoccos*), sometimes called PACIFIC CRANBERRY, is a smaller species than the preceding one. It, too, has curved-back petals, but its leaves are pointed, have rolled edges and are white underneath. Its berries are *bright red.* It is found in swamps and bogs from Newfoundland to Alaska, south to Virginia and northern California.

WOODY VINES AND SHRUBS • NIGHTSHADE FAMILY

MATRIMONY-VINE
Lycium halimifolium

Other names: COMMON MATRIMONY-VINE, EUROPEAN MATRIMONY-VINE

Description: This is a scrambling vine or upright shrub with supple, recurving branches. The stems are ridged and usually somewhat thorny. The leaves are gray-green, narrow or wedge-shaped, with fleshy texture. They are alternate, but usually smaller leaves grow in a cluster in the leaf axils. Small, purplish flowers are replaced by *orange-red,* roundish berries that mature in late summer.

Habitat: This vine grows on waste ground locally from Canada southward. It has escaped from cultivation.

Remarks: Matrimony-Vine is an excellent plant for training over fences and latticework.

Matrimony-Vine

Pale Wolfberry

Berry use: Although the leaves and young shoots are reportedly poisonous to cattle and sheep, the berries are said to be edible raw, cooked or dried.

PALE WOLFBERRY
Lycium pallidum and related spp.

Other names: DESERT THORN, TOMATILLO, SQUAW BERRY, PALE MATRIMONY VINE, WOLFBERRY

Description: These thorny shrubs are stiff and brushy and grow to 6 feet. They are noticeable in winter because their narrow leaves are still present. Their bell-shaped, greenish purple flowers bloom very early. In late summer the shrubs are attractive because of their numerous *red* berries. They sometimes bloom a second time, following summer or early fall rains.

Habitat: Pale Wolfberry or related species grow in the deserts of California, Arizona, New Mexico, Texas, Colorado and Utah.

Berry use: The berries are slightly bitter, juicy or dry, depending on the species. Although they can be eaten raw, they are best made into sauce. They have played an important role in providing food for the Indians in the areas where they grow. Birds also enjoy them and use the shrubs for roosts at night.

WOODY VINES AND SHRUBS • HONEYSUCKLE FAMILY

HONEYSUCKLES
Lonicera spp.

Honeysuckles are erect shrubs or vines with scales remaining at the bases of the twigs. The leaves are simple and opposite. The tubular flowers have an upper and lower lip or a trumpet shape. They are borne in end clusters or on stems springing from the leaf axils. The berries, about ¼ inch in diameter, are juicy, have several seeds and are eaten by numerous kinds of birds. Humans find them edible, but some are not palatable eaten raw and can be considered only survival food. Northern Fly Honeysuckle and American Fly Honeysuckle are two species that are used for jelly or jam, with pectin added. One species or another can be found in most regions of the United States. Here are brief descriptions of some honeysuckles with red berries.

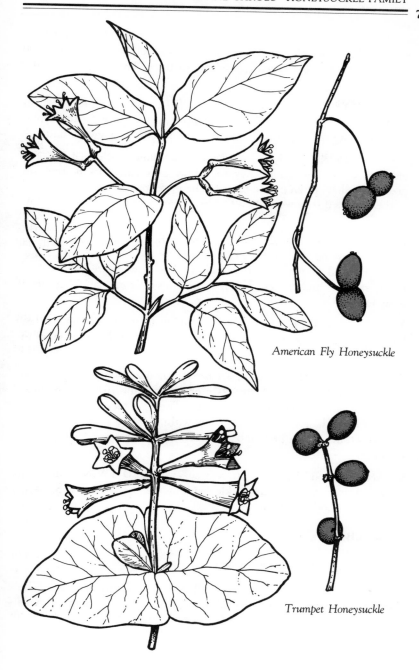

American Fly Honeysuckle

Trumpet Honeysuckle

AMERICAN FLY HONEYSUCKLE (*L. canadensis*) grows 2 to 4 feet high in moist, cool woods. Its flowers are greenish yellow. The berries are waxy, *bright red*, egg-shaped, paired but distinct. They ripen from July to September.

SWAMP FLY HONEYSUCKLE (*L. oblongifolia*) grows 2 to 5 feet high in cold swamps and bogs. Its flowers are white, often tinged with purple. The fruits are *red or purplish*, egg-shaped and often somewhat united.

SMOOTH HONEYSUCKLE (*L. dioica*), sometimes called GLAUCOUS HONEYSUCKLE, has a tendency to sprawl or recline. Its flowers are greenish yellow, often purple-tinged. The upper pair of leaves are joined at the base as if they were one. The berries are *salmon red*, round and in clusters.

ROCK HONEYSUCKLE (*L. prolifera*), also called GRAPE HONEYSUCKLE, is a low-climbing vine with its two upper leaves joined into one oblong leaf, whitened above as well as below. The flowers are yellow in a nearly stalkless end cluster. The round berries are *coral red*. They ripen from July to October.

YELLOW HONEYSUCKLE (*L. flava*) is similar to the preceding species, but the leaves are grayish underneath. The round *red* berries grow in clusters.

TRUMPET HONEYSUCKLE (*L. sempervirens*), also called CORAL HONEYSUCKLE or FIRECRACKER VINE, is a climbing shrub that is sometimes deciduous and sometimes evergreen. The spectacular flowers are trumpet-shaped, scarlet outside, yellow within, 1½ to 2 inches long, in several whorls on an end stalk. The *scarlet* berries ripen from July to October.

ORANGE HONEYSUCKLE (*L. ciliosa*) and RED TWINBERRY (*L. utahensis*) are western Honeysuckles with *orange or red* berries. See also Blue Berries and Black Berries.

CORALBERRY
Symphoricarpos orbiculatus

Other names: INDIAN-CURRANT

Description: This shrub, 2 to 6 feet tall, has fine, downy branches. The short-stalked leaves are egg-shaped, gray-green and smooth above, with

Coralberry

fine down beneath and margins that are sometimes wavy-edged. Bell-shaped flowers, usually pink, grow in clusters at the leaf axils. The small, clustered berries are *coral red to purple*. They ripen from September to November and last into winter.

Habitat: Coralberry grows in low woods or dry, rocky slopes from Massachusetts to Colorado and south to Texas and Georgia. It has escaped cultivation.

Remarks: This shrub is frequently cultivated as an ornamental.

Berry use: Although game birds, such as ruffed grouse and pheasants, occasionally eat the berries, most birds do not care for them. Humans regard them as inedible.

HIGHBUSH-CRANBERRY
Viburnum trilobum

Other names: CRANBERRY VIBURNUM, SQUAW-BUSH, CRAMP-BARK

Description: This is a tall shrub, 3 to 12 feet, with 3-lobed maplelike leaves and a sweet-sour smell. It bears two kinds of white flowers in the flower cluster, the marginal ones large and showy, but sterile, the inner ones small. The fruit is a drupe, about ⅜ inch in diameter, round or slightly oval, a translucent *bright red*. It ripens from September to October.

Habitat: Highbush-Cranberry grows in cool, moist woods and wetland areas from Newfoundland to British Columbia, south to New Jersey, Iowa and Oregon.

Remarks: This shrub is not related to true cranberries. It is often planted as an ornamental.

Use of berrylike fruit: The fruits are hard but soften after the first frost. They are juicy and tart and may be used like cranberries, though you may want to strain out the seeds.

Similar species: SQUASHBERRY — see Yellow or Orange Berries. For other viburnum species, some of which have red fruits at one stage of development, see Black Berries.

Highbush-Cranberry

RED-BERRIED ELDER
Sambucus pubens

This shrub is much like Common Elderberry, with which it should not be confused since, according to some authorities, the berries of Red-Berried Elder are *poisonous* even when cooked. The differences to look for are as follows: The pith of Common Elderberry is white, that of Red-Berried Elder brown. The berries of Red-Berried Elder remain a *brilliant red* even when fully ripe, and the flower and fruit clusters are cone-shaped rather than flat-topped. Red-Berried Elder grows from British Columbia to Newfoundland, south to Georgia, Iowa, Colorado and California. See Common Elderberry in Black Berries.

TREES • YEW FAMILY

PACIFIC YEW
Taxus brevifolia

Other names: WESTERN YEW

Description: This evergreen, 20 to 50 feet tall, has a trunk that is often fluted or malformed and branches that are spreading and almost horizontal. Its needles are flat and linear. They are ½ to 1 inch long, dark green above, light green below, arranged spirally. The berrylike fruit consists of a green seed contained in a *red*, gelatinous cap (aril), with the tip of the seed exposed.

Habitat: The Pacific Yew grows in small groups or as an understory tree in Northwest forests, on deep, moist soil. Near the timberline it is a sprawling shrub.

Use of berrylike fruit: The arils are edible, but the seeds are *poisonous.*

TREES • ELM FAMILY

SUGARBERRY
Celtis laevigata

Other names: SUGAR HACKBERRY, LOWLAND HACKBERRY, SOUTHERN HACKBERRY

Description: This tree is similar to Common Hackberry—see Purple Berries. However, it is a taller tree (60—80 feet high) and grows farther

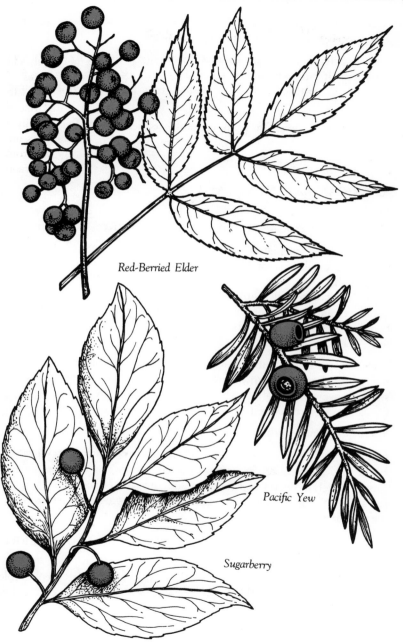

Red-Berried Elder

Pacific Yew

Sugarberry

south. Its berrylike fruits are about ¼ inch long, roundish and typically *orange-red.* They ripen in September or October.

Habitat: Stream banks and swampy places are preferred by the Sugar-berry tree. It ranges from Virginia to Florida, west to New Mexico and northward, in the Mississippi Valley, to Oklahoma, Indiana and Illinois.

Use of berrylike fruit: The fruits are edible either raw or dried.

TREES • ROSE FAMILY

AMERICAN MOUNTAIN-ASH

Scorbus americana

Other names: AMERICAN ROWAN TREE, DOGBERRY

Description: When in fruit, this is a very handsome tree. Rarely more than 30 feet tall, it is sometimes shrub-sized. Its bark is smooth, grayish to reddish-brown, and its branches are slender and spreading. The leaves are compound, with 11 to 17 long, narrow, toothed leaflets. Small, creamy-white flowers bloom in showy clusters. The fruits are berrylike pomes, about ¼ inch in diameter, borne in flat-topped clusters. They are bright *orange-red or orange* at maturity.

Habitat: American Mountain-Ash likes moist soil but also grows well on rocky hillsides. It is often planted as an ornamental. There are several similar species, but all of them like the cool climate of southern Canada, northern United States, and mountainous areas farther south.

Remarks: In spite of the name, Mountain-Ashes are not true ash trees.

Use of berrylike fruit: The fruit, which ripens from August to October, has a very bitter taste, but after repeated freezings it becomes merely tart. It is rich in pectin and can be made into jelly or cooked with sugar for sauce. Songbirds such as evening and pine grosbeaks, cedar wax-wings and robins are fond of the fruit.

HAWTHORNS

Crataegus spp.

Other names: THORNAPPLE, THORN PEAR, THORN PLUM, MAY TREE, HAW

American Mountain-Ash

Hawthorn

Description: The Hawthorns include many species of dense shrubs and small trees. They are distinctive as a group, but the various species are extremely difficult to determine. Typically, they have stiff, sharply pointed thorns on their stems and branches. The smooth bark breaks up into scaly plates with age. The leaves are alternate, simple, conspicuously toothed or lobed. The 5-petaled white, pink or red flowers are in showy terminal clusters which appear early in the spring. The fruits are pomes, generally less than ½ inch in diameter, dry and mealy. They are *usually red,* but may be yellow, orange, blue, purple or black. They ripen in October and may remain on the tree through the winter. Sometimes called haws, they contain between 1 and 5 bony nutlets and have a leathery skin.

Habitat: Abundant from coast to coast in the United States and Canada, Hawthorns sometimes are cultivated because of their showy blossoms and attractive fruits.

Remarks: Hawthorns can readily be distinguished even in winter because their long (1 to 5 inches), sometimes curved thorns stand out clearly.

Use of berrylike fruit: All hawthorn fruits are edible, but some are dry and others don't have a good flavor. They can be steeped into a tea or made into jelly.

TREES • HOLLY FAMILY

AMERICAN HOLLY
Ilex opaca

Description: This is a tree that can grow as high as 100 feet. The short branches form a pyramid-shaped crown of leathery evergreen leaves with spiny teeth. Small flowers are followed by *bright-red,* berrylike drupes that persist into winter. Holly branches stay fresh-looking long after being cut, making them ideal for decoration.

Habitat: American Holly grows in eastern and south central United States, in dry soils at the forest edges or rich, moist swamps.

Remarks: Colonists cultivated and pruned American Holly as a garden hedge; it was especially effective because of its prickly leaves.

Use of berrylike fruit: Poisonous! Even a few fruits can cause violent

American Holly

Saguaro

vomiting in humans. Some of our favorite birds, including bluebirds, kingbirds and hermit thrushes, seem able to eat them without ill effects.

TREES • CACTUS FAMILY

SAGUARO
Cereus giganteus

Description: This tree, which grows 25 to 50 feet tall, is symbolic of the desert. Sparse branches curve upward and parallel the main trunk. All are columnar and ribbed. The leaves are reduced to gray spines located on the prominent ridges. White flowers open at night and emit a melonlike odor. The fruits mature in July. The berries are egg-shaped, 2 to 3½ inches long, *green but becoming red.* When ripe, they burst open, revealing crimson, juicy pulp, and are sometimes mistaken for red flowers.

Habitat: The Saguaro grows only in very dry areas in Arizona, New Mexico, southern California and the Sonora desert in Mexico.

Remarks: Arizona's state flower is the Saguaro. The plant is protected by state and federal government. The oldest specimens are estimated to be 150 to 200 years old. The Saguaro can survive three years without rain. In dry periods, its trunk shrinks and becomes wrinkled. Woodpeckers and owls nest in holes in the branches.

Berry use: The berries can be eaten raw or made into syrup or wine. The Papago Indians have a festival when the fruit is harvested. They eat the fruit raw, make syrup or wine from it or dry it.

TREES • DOGWOOD FAMILY

FLOWERING DOGWOOD
Cornus florida

Other names: BITTER RED BERRY, BOXWOOD, COMMON DOG-WOOD, CORNEL, DOG TREE

Description: This tree has the most beautiful flowers of all the dogwoods. It grows to 40 feet high. Its leaves are oval with a pointed tip. Twigs and branchlets may be green or dark purple. The bark of the trunk is dark and deeply checkered. Small, greenish white flowers are arranged

Flowering Dogwood

in compact heads surrounded by 4 petal-like bracts that are usually white but may be pink or rose. Each cluster has the appearance of a single flower. The berrylike drupes form in a cluster that ripens from August to November. They are *bright red*, egg-shaped and black-tipped.

Habitat: Flowering Dogwood prefers rich, moist soil. Its range is roughly the eastern third of the United States, with the exception of southern Florida. It is often cultivated.

Remarks: A scarlet dye can be obtained from the roots of this tree.

Use of berrylike fruit: Although commonly regarded as inedible, the berries can be eaten cooked. Caution is urged in eating them raw. The fruits are enjoyed by numerous birds and animals, such as chipmunks, foxes and skunks.

Similar species: PACIFIC DOGWOOD (*C. nuttallii*), also called MOUNTAIN DOGWOOD and WESTERN FLOWERING DOGWOOD, grows as an understory tree in coniferous forests. The flowers, with their showy, petal-like bracts, form clusters up to 6 inches across, making it one of the most conspicuous flowering plants in the West. Blooming occurs in the spring and fall. Fruits from the first bloom may be turning *red* as the second blooming period occurs. Pacific Dogwood ranges from British Columbia to California, east to Idaho. The fruits are edible cooked but are not commonly eaten.

PACIFIC MADRONE
(Heath Family) See Yellow or Orange Berries.

Pacific Dogwood

Dark blue and blue-black are often hard to distinguish, and sometimes blue has a reddish or purplish cast, so check other sections if necessary.

FLOWERING HERBS • LILY FAMILY

CARRION-FLOWER

See Black Berries.

SOLOMON'S-SEAL

(*Polygonatum biflorum*)

Other names: SMALL SOLOMON'S-SEAL, SMOOTH SOLOMON'S-SEAL

Description: This is a perennial plant with a long, usually arching stem, up to 3 feet in height. The leaves are oval, 2 to 4 inches long, pointed at the tip and arranged alternately. The inconspicuous greenish yellow, tassel-like flowers grow in pairs that hang drooping beneath the leaves. The round berries are *dark blue or blue-black* and pulpy, about ¼ inch in diameter. They form in midsummer.

Habitat: This is a common plant in woods and thickets from Maine to the Great Lakes area, south to Florida and Texas.

Remarks: The name comes from ancient Greek writers who observed that the raised, circular scars of former stems on the rootstock looked like the stamp of a seal. The plant can be distinguished from False Solomon's-Seal by the fact that its flowers and berries hang from the leaf axils instead of being located in terminal clusters.

Berry use: The berries of this and the following species are not edible.

Solomon's-Seal

Similar species: GREAT SOLOMON'S-SEAL (*P. canaliculatum* or *commutatum*) is taller and coarser. Its flowers are in clusters of 2 to 10. It is found from southern Manitoba to New England and south to Georgia and Texas. HAIRY SOLOMON'S-SEAL (*P. pubescens*) is smaller than Solomon's-Seal, has smaller flowers and leaves with hairy veins underneath. It is found in southern Canada and northern United States, south to the mountains in South Carolina.

CLINTONIA
Clintonia borealis

Other names: CORN-LILY, YELLOW CLINTONIA, BLUEBEAD, YELLOW BEADLILY

Description: This perennial grows 6 to 16 inches tall, with shining, prominently veined leaves rising from the base of the plant. There are usually 2 or 3 such leaves, 5 to 8 inches long, shaped much like those of Lily-of-the-Valley. The flowers are yellowish green bells hanging in a cluster from a common center on a leafless stalk, each ¾ to 1 inch long. The berries are a beautiful *deep blue*, round, about ¼ inch in diameter. They branch out erect from the top of the leafless stalk. The number of flowers determines the number of berries, which ripen in September.

Habitat: This species is found in woods, thickets and swampy places from Newfoundland to Manitoba, south to Minnesota, Indiana and the mountains of Georgia.

Remarks: The name honors DeWitt Clinton, once governor of New York. The young leaves of Clintonia have a taste similar to that of cucumber and are used for salads or can be boiled and served as a vegetable. Chippewa women chewed patterns in the leaves of Clintonia to make decorative ornaments. Menominee Indians placed the leaves over dog bites in an effort to aid healing.

Berry use: The berries are inedible.

Similar species: RED CLINTONIA (*C. andrewsiana*), also called ANDREW'S CLINTONIA, has red or reddish lavender flowers and *deep-blue* berries. It grows in shaded damp forests near the coast from central California to southwest Oregon. It is one of the few wildflowers that grow in the

Indian Cucumber-Root

Clintonia

redwood forests of the Pacific Coast. QUEENCUP (*C. uniflora*), also called BRIDE'S BONNET, has white, starlike flowers that are solitary or in pairs. The 6 petal-like segments form a broad bell. The berry is *blue*, round or pear-shaped. This species grows on hills and mountains from Alaska to California, east to Idaho and Montana. WHITE CLINTONIA (*C. umbellulata*) carries its flowers erect and has *black* berries.

INDIAN CUCUMBER-ROOT
Medeola virginiana

Description: This plant has a slender, unbranched erect stem with two whorls of oblong, pointed leaves. It grows from 1 to 2½ feet tall and is woolly. The leaves of the lower whorl are 2 to 5 inches long. The leaves composing the upper whorl are smaller. Drooping, greenish yellow flowers have their 6 tips recurved. The berries are *dark blue or purple*, about ¼ to ½ inch in diameter.

Habitat: Indian Cucumber-Root is found in moist woods and thickets from Nova Scotia to Minnesota, south to Florida and Louisiana.

Remarks: The tubers resemble cucumbers in flavor and are sometimes used for salads or pickles. They should not be used where plants are scarce.

Berry use: The berries are inedible.

BLUE COHOSH
Caulophyllum thalictroides

Other names: PAPOOSEROOT, SQUAW ROOT, BLUE GINSENG

Description: This plant grows from 1 to 3 feet high and produces flowers before the leaves are fully developed. The stem grows erect from a perennial, horizontal rootstock, which is poisonous. When the plant is mature, it has a bushy appearance. The compound leaves have wide, thick leaflets, each with 3 to 5 lobes. The developing stems and leaves are often purplish in the spring. Greenish flowers, with 6 petal-like sepals, bloom in a terminal cluster. The berrylike fruits are *blue*, in a loose cluster.

Blue Cohosh

Habitat: Blue Cohosh grows in low woods, moist banks and clearings from New Brunswick to Manitoba south to South Carolina and Missouri.

Remarks: The roots of this plant were once used for medicinal purposes. The word "cohosh" comes from the Algonquian word for rough.

Use of berrylike fruit: The bitter tasting fruits are *poisonous.* It is said, however, that roasted seeds make a safe substitute for coffee.

WOODY VINES AND SHRUBS • CEDAR or CYPRESS FAMILY

COMMON JUNIPER
Juniperus communis

Other names: DWARF JUNIPER, GROUND JUNIPER, SCRUB JUNIPER

Description: This is a flattened evergreen shrub or, rarely, a small tree. The reddish brown bark shreds easily. The needles are sharp, hollow, 3-sided, in whorls of 3. They are grooved and whitened above and are ½ to ¾ inch long. The male flowers open in spring as tiny bunches of yellow blossoms near the twig tips. Female flowers are also very small and look like cones. The aromatic, *blue, purple or black* fruits are often called juniper berries. They are whitened with a bloom, round or egg-shaped, about ¼ inch in diameter, with 1 to 12 seeds.

Habitat: This species grows in exposed rocky or sandy places from New-foundland to Alaska, south throughout Canada and the United States wherever conditions are right, especially on hillsides that are otherwise bare of trees or shrubs.

Remarks: The fleshy fruits are actually formed of flower scales that grow together, along with the scaly bracts below. They take three years to ripen in cool climates, two years in warmer climates.

Use of berrylike fruit: The flavor of gin comes from the fruit of Common Juniper. Crushed berries are sometimes used to season meat, especially veal or lamb. A few berries, 5 or 6, are sufficient for this purpose, and they should be removed before serving the food. They have a mild diuretic effect. Heated berries were once applied to wounds.

Common Juniper

CREEPING JUNIPER
Juniperus horizontalis

Other names: TRAILING JUNIPER

Description: This is a prostrate or creeping evergreen shrub that forms mats. It usually has long, trailing branches with numerous short branchlets. Its leaves are scalelike, bluish green and small, with pointed tips. The berrylike fruits are *light blue,* ¼ to ⅜ inch in diameter, and are borne on backward-curving short stalks.

Habitat: Creeping Juniper thrives on sandy or rocky ground, frequently on the borders of swamps and bogs, from Newfoundland to Alaska, south to northern New England, New York, the Great Lakes area, Nebraska and Wyoming.

Use of berrylike fruit: The fruits are used as seasoning. See Common Juniper. They are also eaten by moose, white-tailed deer and sharp-tailed grouse.

GLAUCOUS GREENBRIER
(Lily Family) See Black Berries.

COMMON MOONSEED
(Moonseed Family) See Black Berries.

WOODY VINES AND SHRUBS • BARBERRY FAMILY

BARBERRIES

The barberry family is a confusing one because it includes two very different groups — evergreens with hollylike leaflets and shrubs with oval, deciduous leaves. The former are often called MAHONIA. See Red Berries for deciduous barberries. Barberries with blue berries include the following:

OREGON GRAPE
Berberis repens

Other names: CREEPING OREGON GRAPE, HOLLY GRAPE, CREEPING MAHONIA, CREEPING BARBERRY

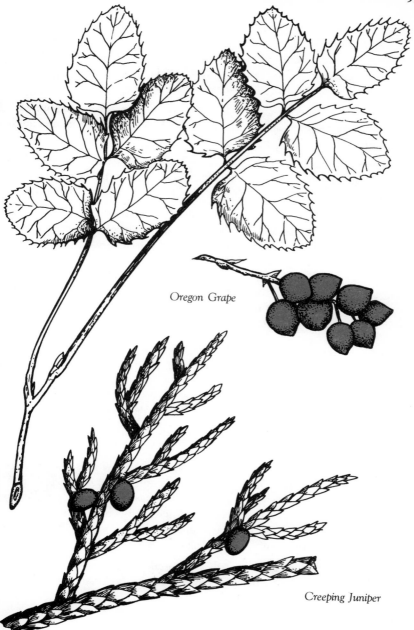

Oregon Grape

Creeping Juniper

Description: This shrub has 2 or 3 compound leaves with leathery, evergreen, spiny-margined leaflets on low, short stems. It spreads by underground stems and grows to 8 inches tall. The flowers, about ½ inch in diameter, are yellow, bunched in dense, branching clusters. They are followed in autumn by *chalky blue* berries about ¼ inch wide.

Habitat: This species can be found in open pine forests and wooded areas from western Canada south to California and Texas and north to South Dakota.

Berry use: The berries are sour, but they can be eaten raw; they also make good jelly, pie or wine. They are enjoyed by many forms of wildlife, including bears.

Similar species: CASCADE OREGON GRAPE (*B. nervosa*) has more leaflets and is a shrubby species. The berries are *dark blue to purplish blue.* TALL OREGON GRAPE (*B. aquifolium*), also called OREGON HOLLYGRAPE, OREGON GRAPE and HOLLY BARBERRY, is the state flower of Oregon. It has hollylike leaves and *dark-blue* berries in grapelike clusters. The berries become sweeter after the first frost and make delicious jelly. Species found in the deserts of the Southwest include *B. haematocarpa*, *B. fremontii* and *B. trifoliolata*.

JUNEBERRIES
See Purple Berries.

WOODY VINES AND SHRUBS • BUCKTHORN FAMILY

SUPPLEJACK
Berchemia scandens

Description: This high-climbing, woody, deciduous vine is a member of the buckthorn family. Its alternate leaves are elliptic or lance-shaped. They have distinctively straight, prominent side veins and are slightly paler underneath. The flowers are small, greenish white, growing in end clusters. The berrylike fruits are oval-shaped, *blue or bluish black,* each with 1 seed. They mature from August to October.

Habitat: Supplejack grows chiefly in coastal plains from Virginia south to Georgia and west to Texas. It grows north in the Mississippi Valley to Missouri.

Supplejack

Use of berrylike fruit: The berries are regarded as inedible by humans, but they are eaten by numerous birds, including wood ducks, mallards, bobwhites and wild turkeys.

WOODY VINES AND SHRUBS • VINE FAMILY

HEARTLEAF AMPELOPSIS
Ampelopsis cordata

Other names: AMERICAN AMPELOPSIS

Description: Similar to Muscadine Grape, this climbing vine has a white pith rather than a brown one. Tendrils may be present opposite some of the leaves. The leaves are broadly egg-shaped, pointed, with the base heart-shaped or flattened. They are coarsely and irregularly toothed, sometimes 3-lobed, smooth above and sometimes slightly downy underneath. The small, greenish flowers grow in long-stalked, forking clusters opposite some of the leaves. *Blue* berries, pea-sized, usually 2-seeded, ripen from August to November.

Habitat: This vine grows in swamps and fertile woods from Maryland, West Virginia, Illinois, Missouri and Oklahoma south to Florida and Texas.

Berry use: Humans find the berries dry and inedible, but several songbirds and bobwhites enjoy them.

VIRGINIA CREEPER
Parthenocissus quinquefolia

Other names: WOODBINE, AMERICAN IVY, FIVE-LEAVED IVY

Description: This is a common climbing vine with 5 (sometimes 3 or 7) leaflets arranged like an open fan from the end of a long leafstalk. The leaves are dull above, paler and sometimes hairy underneath. The tendrils branch out, each one ending in a small adhesive disc. Small, greenish flowers are clustered at the ends of the branches. The berries ripen from August to October. They are round, *dark blue,* not very fleshy, about ¼ inch in diameter, with 2 or 3 seeds.

Habitat: Virginia Creeper thrives in woods, thickets, on rocky banks from Quebec to Minnesota and Colorado, south to Florida and Texas. It is often planted as a climber on porches, trellises and walls.

Heartleaf Ampelopsis

Remarks: This vine is sometimes mistaken for Poison Ivy. However, it usually has 5 leaflets, compared to Poison Ivy's 3. Poison Ivy berries are white; those of Virginia Creeper are dark blue.

Berry use: Although some people eat the berries, they are *poisonous* and can prove fatal if eaten in sufficient quantity.

Similar species: THICKET CREEPER (*P. inserta*) is very similar, but the tendrils only rarely end in adhesive discs. The leaves are dark green and shiny above. It grows in the same area, but is also found in Arizona and California. The berries are *poisonous.*

WILD GRAPES

See Purple Berries.

WOODY VINES AND SHRUBS • DOGWOOD FAMILY

DOGWOODS

Dogwoods are usually shrubs, but some grow to the size of a small tree. The name comes from the fact that the wood was used for "dogs," or skewers, in the Middle Ages. With one exception, dogwoods have opposite, simple leaves, without teeth. Leaf veins curve and tend to parallel the margins of the leaf and meet at its tip. The flowers are small, greenish yellow or white and borne in terminal clusters. The fruits have a large stone enclosing the seed and are surrounded by a fleshy portion. They have the appearance of berries. There are about 100 species of dogwood. The most common species with blue berries are as follows:

SILKY DOGWOOD

Cornus amomum

Other names: KINNIKINNICK, SWAMP DOGWOOD, PALE DOGWOOD

Description: This shrub, 4 to 10 feet tall, has branchlets that are purplish red, covered with silky down. The leaves are 2 to 4 inches long, egg-shaped, pointed, rounded at the base. They are smooth above and usually have some reddish hairs beneath. Creamy-white flowers bloom in flat-topped clusters. The berrylike fruits are round, *dull blue or partly white,* about ¼ inch in diameter. They ripen from August to October.

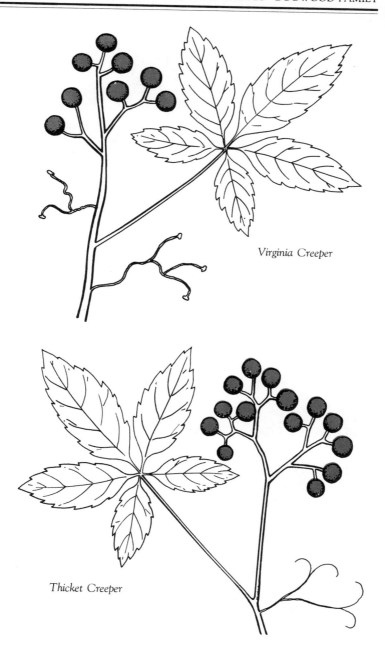

Virginia Creeper

Thicket Creeper

Habitat: Silky Dogwoods grow in wet or moist places from southern Maine to Minnesota, south to Georgia and Oklahoma.

Remarks: A mixture of the bark and dried leaves of this shrub, sometimes mixed with tobacco, was smoked by the Indians and called kinnikinnick.

Use of berrylike fruit: Although dogwood fruits in general can be eaten raw or cooked, they have a laxative action when eaten raw, and are regarded as a survival food rather than a desirable fruit. They are apt to taste very bitter.

ROUNDLEAF DOGWOOD
Cornus rugosa

Description: The distinguishing mark of this dogwood is that the twigs are greenish or reddish brown, marked with purplish blotches. The flowers are white; the fruits are *pale blue, occasionally white.*

Habitat: This species grows from Nova Scotia to Manitoba, south to New England, Virginia, the Great Lakes region and northeastern Iowa.

Remarks: The twigs are eaten by white-tailed deer, moose and cottontails.

Use of berrylike fruit: The fruits can be eaten cooked, but caution is advised for raw fruits. Ruffed and sharp-tailed grouse are fond of the fruits.

ALTERNATE-LEAF DOGWOOD
Cornus alternifolia

Other names: BLUE DOGWOOD, PAGODA DOGWOOD

Description: This dogwood can be told from the others because its leaves are alternate rather than opposite. It can be a shrub or it can grow to 30 feet tall. The flowers are creamy white. The berries, borne on bright-red stalks, are *deep blue to black.*

Habitat: This species likes moist woods and stream banks from Newfoundland to Manitoba and Minnesota, south to Florida and Arkansas.

Use of berrylike fruit: See preceding. Many birds, including ruffed grouse, eat the fruits.

Silky Dogwood

Roundleaf Dogwood

Alternate-Leaf Dogwood

POISON DOGWOOD

This should properly be called Poison Sumac, since it does not belong to the dogwood family. See Poison Sumac, White or Green Berries.

BLUEBERRIES
Vaccinium spp.

Blueberries, huckleberries, bilberries and deerberries often closely resemble one another and may hybridize, making identification of species extremely difficult. Blueberries differ from huckleberries, bilberries and deerberries in that their twigs are densely covered with fine, warty speckles. See Huckleberries for other differences between blueberries and huckleberries.

Description: There are several species of wild blueberries. In general, they are characterized by small, short-stalked, elliptic simple leaves arranged alternately. The twigs are slender, green or reddish, sometimes zigzag, with numerous speckles or "warts" raised above the surface. Height varies from 8 inches to 15 feet tall or more. The small, attractive, bell-shaped flowers are white, pinkish or greenish. The berries are round, with a star-shaped pattern on the upper tip, formed by 5 calyx lobes. Their color is *usually blue, but may be red, purple or black*, often with a white powder. Blueberries ripen from June through September.

Habitat: Blueberries require an acid soil, and thrive especially well on fire-blackened land. They are usually found in open woods and clearings. Depending on the species, they may also grow in bogs, tundras and barrens. They are found in the United States wherever conditions are favorable, and the same is true of huckleberries and bilberries.

Remarks: Blueberries seem to require the growth of a root fungus called mycorrhiza, which also thrives in acid soil.

Berry use: All species of this sweet, juicy berry are edible, though some are tastier than others. HIGHBUSH BLUEBERRY (*V. corymbosum*), which grows in the East, is reputed to take the prize for sweetness. Blueberries are used fresh, cooked, dried, in pies and, with pectin added, in jellies or jams. Birds and mammals also enjoy them. See also Black Berries.

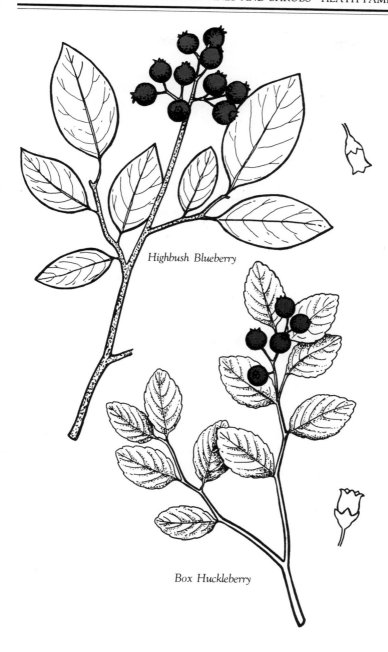

Highbush Blueberry

Box Huckleberry

HUCKLEBERRIES
Gaylussacia spp.

Huckleberries are much like blueberries except that (1) huckleberry twigs are not covered by "warts"; (2) except for Box Huckleberry, huckleberry leaves are dotted with small, glistening yellow globules of resin; (3) the berries are mealier and have 10 fairly large seeds, instead of many small ones. Species of huckleberries with blue berries include TALL HUCKLEBERRY (G. *frondosa*), growing to 6 feet, with berries that are *dark blue* with white powder. Tall Huckleberry is also called DANGLEBERRY, TANGLEBERRY, BLUE-TANGLE, or BLUE HUCKLEBERRY. BOX HUCKLEBERRY (G. *brachycera*) is a low evergreen species with *blue* berries that have a whitish bloom. See also Black Berries.

BILBERRIES
Vaccinium spp.

Like huckleberries and deerberries, bilberries can be distinguished from blueberries by the lack of warty speckles on the twigs. Also, the berries arise from leaf axils instead of being in terminal clusters. They are divided into several species: DWARF BILBERRY (V. *caespitosum*) has berries that are a *light, rather than deep,* blue. OVAL-LEAF BILBERRY (V. *ovalifolium*) is a straggling, slender shrub with *blue* berries whitened with a bloom. See also Black Berries.

DEERBERRIES
See White and Green Berries.

NORTHERN HONEYSUCKLE
Lonicera villosa

Other names: MOUNTAIN FLY HONEYSUCKLE, BLUE HONEYSUCKLE, NORTHERN FLY HONEYSUCKLE

Description: This is a small (1 to 3 feet high), erect shrub with a solid pith and shredding back. The branchlets may be either hairy or smooth. The blunt leaves are usually hairy on both sides and are paler underneath. The leaf margins are untoothed but are usually fringed with fine hairs. The flowers are pale yellow and tubular, with 5 lobes. They grow

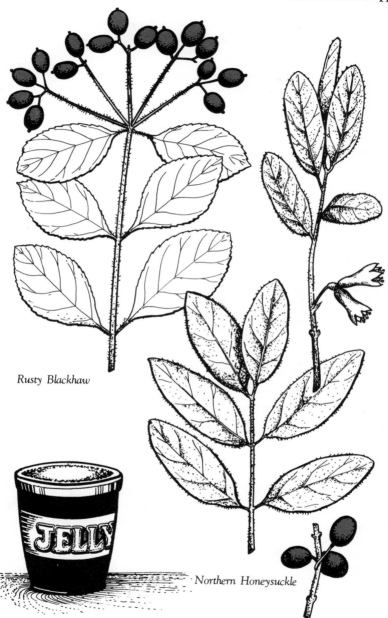

Rusty Blackhaw

Northern Honeysuckle

in pairs on short stalks that arise from the leaf axils. The berries are *dark blue*, oval, two-eyed and about ¼ inch in diameter. They ripen from June to August.

Habitat: Northern Honeysuckle is found in bogs, swamps, rocky or peaty soil from Newfoundland to Manitoba, south to northern United States.

Remarks: It is not commonly known that this berry is an edible one.

Berry use: The berries can be eaten raw, cooked into sauce or used for jelly or jam, with pectin added. See also Red Berries and Black Berries.

RUSTY BLACKHAW
Viburnum rufidulum

Other names: SOUTHERN BLACKHAW, RUSTY NANNYBERRY

Description: This is a shrub or small tree 5 to 25 feet tall. It is similar to BLACKHAW (see Black Berries) except that the undersides of the leaves, the leafstalks, the buds and sometimes the twigs are densely hairy with rusty or reddish hairs. The leafstalks are slightly winged. The oval-shaped berrylike drupes develop from the small, white flowers. They are red when immature, *deep blue when mature,* about ½ inch long.

Habitat: This shrub grows in dry pine or deciduous woods and thickets from Virginia to Kansas, south to Florida and Texas.

Use of berrylike fruit: The fruits are sweet and edible and can be eaten fresh, cooked, or made into jelly. See Black Berries for other viburnum species and for ELDERBERRIES.

EASTERN RED CEDAR
Juniperus virginiana

Other names: CEDAR, SAVIN, RED JUNIPER. The latter name is the most proper (though not the most usual) one, for this tree is really a juniper, not a cedar.

Description: This is a medium-sized tree, usually 30 to 40 feet tall, although it sometimes grows much larger. Young trees have narrow,

Eastern Red Cedar

cone-shaped crowns, but the tree becomes irregular and round-topped as it grows older. The bark is thin, pale reddish brown, and it shreds into long, narrow strips. The branches are densely covered with overlapping, dark-green, scalelike leaves in such a way that the twig appears square in cross-section. Near the top of the branches, the new foliage is prickly and pointed. The fruit resembles a berry, but it is actually a cone with fleshy scales. It is green at first, *blue at maturity*, covered with a white bloom, and is about ¼ inch in diameter.

Habitat: This is a tree of dry and rocky soils. It sometimes invades abandoned fields and is found on hillsides with limestone outcroppings. It ranges throughout most of eastern United States, west to the Dakotas, Texas and New Mexico.

Remarks: The wood of this tree is used for fence posts because of its resistance to rot, for pencils and for "cedar closets," because of its fragrance. Oil of cedar is distilled from the leaves and wood and is used in polishes, medicines and perfumes. Red Cedar acts as an alternate host to apple rust.

Use of berrylike fruit: The fruits are sometimes used for flavoring (see Common Juniper). They are also eaten by opossums and over 50 species of birds.

Similar species: SOUTHERN RED CEDAR (*J. silicicola*) is very similar, but it has very small fruits and slender, pendulous branches. It is found on swampy coastal plains from North Carolina to Texas.

TREES • LAUREL FAMILY

SASSAFRAS
Sassafras albidum

Other names: AGUE TREE, CINNAMON WOOD

Description: This is a small tree or shrub usually 20 to 50 feet high. It grows larger than this in the southern portion of its range. It has a flat, open crown. In summer it is easily recognized by its spicy odor, yellowish green branchlets and variable leaves. The leaves are alternate, simple and deciduous. Three shapes appear on the same tree: 2-lobed (mitten-shaped), 3-lobed and unlobed. Small, yellow-green flowers appear in racemes with the leaves. The berrylike fruits are drupes, each

Sassafras

borne on a bright-red stalk. They are *dark blue* and shiny, pea-sized, with a yellowish base.

Habitat: The presence of this tree indicates poor soil. It grows along fence rows and abandoned fields, often forming dense thickets. Its range is from Massachusetts to Michigan and Kansas, south to Florida and Texas.

Remarks: Oil distilled from the bark is used as a flavoring in candies and medicines and to perfume soap. In former days a "spring tonic" was made by boiling roots or bark, but recent experiments show that the tree contains a chemical that causes cancer in laboratory animals. A bark extract can be used to dye wool orange.

Use of berrylike fruit: The fruits are unpalatable, and it is best to leave them alone. They are used to make wine, however.

REDBAY
Persea borbonia

Other names: SWAMP BAY

Description: This is a small to medium-sized tree, usually growing less than 60 feet tall, with a rounded, dense crown. The leaves are shiny, leathery, evergreen, paler underneath. They are narrowly elliptic, toothless, and smell of bay leaves. From May to July, clusters of small, creamy white flowers bloom in the leaf axils. The berrylike fruit is *dark blue or black,* borne on red stems. The calyx is present.

Habitat: Wet swampy areas or moist sandy woods are home for the Redbay, which grows on the coastal plains from Delaware to Florida and west to Texas.

Remarks: The fresh or dried leaves are the bay leaves we use as seasoning.

Use of berrylike fruit: The fruits can be eaten raw, but are not good.

HAWTHORN
(Rose Family) See Red Berries.

Redbay

TREES • TUPELO FAMILY

BLACK TUPELO
Nyssa sylvatica

Other names: BLACK GUM, SOUR GUM, PEPPERIDGE, SWAMP BLACK GUM

Description: This is a medium-sized tree, 30 to 80 feet tall, which occasionally reaches a height of 100 feet or more. It has numerous spreading and often horizontal branches. The simple, alternate, oval leaves are thick and lustrous, dark-green above, paler and often hairy below. They turn scarlet in the fall. Small, greenish flowers are borne on separate stalks. The berrylike drupes are *blue or blue-black,* about ½ inch long, oval, borne 2 or 3 together on a long, thick red stalk.

Habitat: Moist, rich bottomland, dry mountain ridges, abandoned fields and cold mountain swamps are all favorite habitats of Black Tupelo. It is never abundant. Its range is from Maine to Michigan, south to Florida and Texas.

Remarks: This tree is often planted as an ornamental because of its splendid autumn foliage. Its wood is used for furniture, crates and paper pulp.

Use of berrylike fruit: The fruit is edible but acrid. It is sometimes used for preserves. Many kinds of mammals and birds eat the fruit.

See WATER TUPELO, Purple Berries.

FRINGETREE
(Olive Family) See Purple Berries.

Black Tupelo

Only berries that are white or green at maturity are included here.

FALSE SOLOMON'S-SEAL and STARRY FALSE SOL-OMON'S-SEAL
(Lily Family) See Red Berries.

FLOWERING HERBS • SANDALWOOD FAMILY

PALE COMANDRA
Comandra pallida

This plant is similar to Northern Comandra (see Red Berries), but its leaves are pale gray-green and narrower. Comandras are parasitic on the roots of various woody plants. Pale Comandra grows throughout the West, in pine woods and on the prairie. The *green* drupes are hard but edible. They turn brown in time.

FLOWERING HERBS • BUTTERCUP FAMILY

WHITE BANEBERRY
Actaea pachypoda and *arguta*

Other names: DOLL'S EYES, SNAKEROOT, WHITE COHOSH, SNAKEBERRY, CHINABERRY

Description: This perennial plant has erect stems 1 to 2 feet high. It has a bushy appearance. Its large, compound leaves are divided and sub-divided into sharply toothed leaflets. The small, white flowers have narrow petals and bushy stamens, and they grow in fluffy clusters at the end of a long, naked stem. Maturing from July to October, the attractive *white* berries, about ½ inch long, are clustered on a thick, red stem. They are glossy, oval or round, with a dark spot in the center, which is like the pupil of an eye and results in the name "Doll's Eyes."

Habitat: White Baneberries are found in rich, moist woods and thickets from Quebec to Alaska, south to Georgia and California.

Remarks: The berries look as if they were made of china, hence the name "Chinaberry."

Berry use: All parts of the plant, particularly the roots and the berries, contain a *poisonous* glycoside. Eating even a few berries can cause distress, resulting in vomiting, cramps, delirium, dizziness and circulatory failure. Some birds apparently eat the berries without ill effects.

COMMON STRAWBERRY
(Rose Family) See Red Berries.

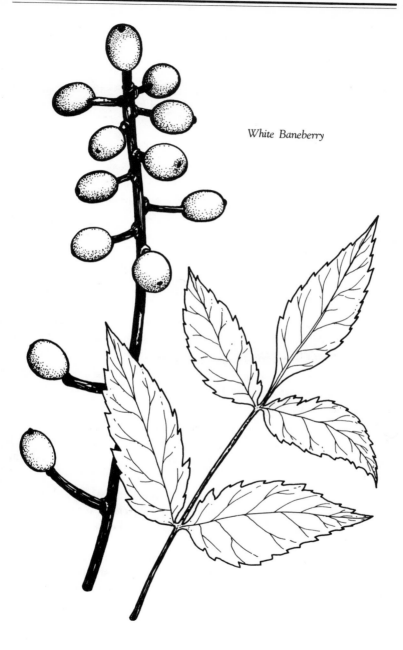

White Baneberry

FLOWERING HERBS • NIGHTSHADE FAMILY

PURPLE NIGHTSHADE
Solanum xanti

This relative of Bitter Nightshade (see Red Berries) differs in that its purple flowers are flat, its leaves are oval and its fruit is a *green, poisonous* berry. It is found in California.

WOODY VINES AND SHRUBS • BAYBERRY FAMILY

SOUTHERN BAYBERRY
Myrica cerifera

Other names: WAX MYRTLE, COMMON WAX MYRTLE, SOUTHERN WAX MYRTLE, CANDLEBERRY

Description: This shrub or small tree, 10 to 30 feet tall, is erect and stiff-branched. Its short-stemmed, alternate, evergreen leaves are narrow, with coarse teeth above the middle. They are yellowish green, with small, resinous dots, and are fragrant when crushed. The fruits, which look like very small berries, are round, light-green drupes, thickly coated with whitish wax.

Habitat: Wet, sandy pine lands and bogs provide the right conditions for this species. It grows on coastal plains from New Jersey south to Florida and west to Texas, and also grows north into Arkansas and Oklahoma.

Use of berrylike fruit: The wax of berries of this and other species in the bayberry family is melted to make bayberry candles. Many birds, including bobwhite quails and wild turkeys, enjoy eating the berries, but humans do not.

Similar species: NORTHERN BAYBERRY (*M. pensylvanica*) is a shrub that has bony fruits covered with minute hairs and *whitish wax*; it grows in the Northeast, inland to the eastern Great Lakes area. EVERGREEN BAYBERRY (*M. heterophylla*) has smooth, *wax-covered* fruits and twigs covered with black hair. It grows on coastal plains from New Jersey to Louisiana. PACIFIC BAYBERRY (*M. californica*) is a shrub or small tree with *dark-purple fruits covered with gray wax*. It grows along the Pacific Coast.

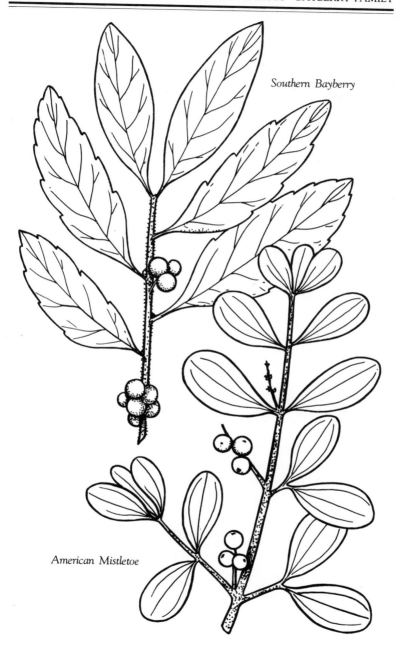

Southern Bayberry

American Mistletoe

WOODY VINES AND SHRUBS • MISTLETOE FAMILY

AMERICAN MISTLETOE
Phoradendron flavescens

Description: This evergreen shrub is yellowish green with numerous jointed branches. It is parasitic on the branches of oaks and other broad-leaved trees. The leaves are thick and leathery, oval or egg-shaped. Small, inconspicuous flowers bloom in October or November. The berrylike fruits, which do not mature until the next autumn, are *waxy white*, transluscent and soft. The pulp is clear but sticky.

Habitat: The range of this parasitic shrub is from New Jersey to Illinois and Kansas, south to Florida and Texas.

Remarks: This plant is spread principally by birds, who eat the fruits and then wipe their sticky bills on tree branches, thus spreading the seeds. It is often used as a Christmas decoration and is the state flower of Oklahoma.

Use of berrylike fruit: The berries are *poisonous* to humans. A number of bird and animal species eat the fruit with no ill effects.

Similar species: See Mistletoe, Red Berries.

WOODY VINES AND SHRUBS • SAXIFRAGE FAMILY

GRANITE GOOSEBERRY
Ribes curvatum

Description: See Purple Berries for a general description and illustrations of gooseberries. This species has branches that recurve, or droop. The branchlets are reddish brown to purplish brown, with short, red spines. The *greenish* berries ripen from June to August.

Habitat: Granite Gooseberries are found in rocky woods and hillsides in northern Georgia, Alabama and Tennessee.

Berry use: The berries can be used for sauce, jelly, jam or pie.

Similar species: FLORIDA GOOSEBERRY (*R. echinellum*) also has *greenish* berries, but they are densely covered with slender, gland-tipped spines. It is found in South Carolina and Florida.

Poison Ivy

WOODY VINES AND SHRUBS • CASHEW FAMILY

POISON IVY

Rhus radicans

Other names: MARKWEED

Description: Poison Ivy can be an erect shrub 2 to 7 feet tall or a vine that trails or climbs on plants or trees. It is commonly described as 3-leaved, but it actually has compound leaves with 3 leaflets. The leaves are alternate and long-stalked. Unfortunately for purposes of recognition, there is great variation in the leaflets. They may be glossy or dull; hairless or slightly hairy; toothless or saw-toothed; leathery or thin; coarse-toothed, wavy-edged or smooth on the margins. Young or dying leaves may be yellow or reddish. The end leaflets have longer stalks than the side ones and have a pointed tip. To identify this plant, remember that Poison Ivy and Poison Oak (equally dangerous) are the only common thornless, woody plants with alternate leaves composed of 3 leaflets.

Small, yellowish flowers, often hidden by the leaves, bloom from May to July. The berrylike drupes, which grow in axillary clusters, are *yellowish white*, waxy, less than ¼ inch in diameter and round. They mature from August to October and persist. Those who do not recognize the plant have been known to pick them for winter bouquets.

Habitat: Poison Ivy thrives on ground that has been disturbed, such as roadsides, but may also be found in woods and thickets. A native plant, it is found in nearly every part of Canada and the United States, with the exception of California.

Remarks: As almost everyone knows, Poison Ivy contains a dangerous skin irritant. Touching the leaves, berries, stalks or roots can have very bad effects. Smoke that contains suspended particles of the plant has been known to cause a skin rash. Even touching an animal or a piece of clothing that has come in contact with the plant can result in a rash. So can breaking the stems in midwinter. If you accidentally touch Poison Ivy, washing with strong soap and water or juice of the Jewelweed plant (*Impatiens spp.*) within a few hours may prevent a rash.

Use of berrylike fruit: DO NOT TOUCH THESE FRUITS or any other part of the plant. You can get an internal rash from eating the berries. Al-

Poison Oak

Poison Sumac

though the berries have poisonous effects on man, over 60 species of birds, including ruffed and sharptail grouse, bobwhite quails and prairie chickens, eat them without ill effects. The seeds pass through their digestive systems and in this way the distribution of the plant is increased.

Similar species: POISON OAK *(R. toxicodendron)* has flowers and fruits that are similar to those of Poison Ivy, and it is found in California as well as in most other states. However, Poison Oak is always a stiffly erect shrub and the fruits are usually hairy. The 3 leaflets are downy and often have 3 to 7 lobes, resembling oak leaves. The skin irritation resulting from touching this plant is much the same as that caused by Poison Ivy.

POISON SUMAC
Rhus vernix

Other names: POISON ELDER, POISON DOGWOOD

Description: DO NOT TOUCH THIS PLANT. All parts of it contain a dangerous skin irritant. The poison is more virulent than that of Poison Ivy and the symptoms are similar. Although usually a shrub, it can grow to 30 feet. Its twigs are hairless and the number of leaflets is less than that of other sumacs, only 7 to 13. They are not toothed. The flowers are small and greenish, in clusters. The berrylike fruits are round, waxy, *ivory-white or greenish white.* They resemble Poison Ivy berries and persist into winter.

Habitat: Fortunately, this species grows primarily in swamps and tamarack bogs, from Maine to Ontario and Minnesota south to Florida and Texas.

Use of berrylike fruit: The fruit is *poisonous* for humans but is eaten by several kinds of birds, including bobwhite quails, pheasants and ruffed grouse.

WOODY VINES AND SHRUBS • OLEASTER FAMILY

SILVERYBERRY
Elaeagnus commutata

Other names: AMERICAN SILVERYBERRY

Silveryberry

Description: This thornless, much-branched shrub is noted for its distinctively rusty and silvery scales. It grows to 12 feet. The leaves are egg-shaped, silvery brown and scaly on both sides. They may be either smooth or wavy-edged. Fragrant, silvery yellow flowers appear in June or July. The berrylike fruits ripen from July to October. They are small, fleshy and *silvery* in color, with tough skins and a single seed.

Habitat: Silveryberries grow in dry soil from Quebec to Alaska, south to Utah and Minnesota.

Remarks: The only other plant with silver-brown scales is Buffaloberry, which is related. The elaeagnus species is among the few non-legume plants that fixes nitrogen in the soil by the action of bacterial root nodules.

Use of berrylike fruit: The berries are dry but may be eaten raw or cooked in pie or jelly.

Similar species: RUSSIAN OLIVE (*E. angustifolia*) is an introduced shrub or small tree that sometimes escapes cultivation. It has silvery leaves and branches. The fruit is an edible, *silvery*, berrylike drupe.

WOODY VINES AND SHRUBS • DOGWOOD FAMILY

RED-OSIER DOGWOOD
Cornus stolonifera

Other names: This is one of the species sometimes called KINNIKINNICK.

Description: This shrub or small tree is notable for its branchlets, which are blood red, a lovely sight in a snowy landscape. Its leaves are egg-shaped, smooth or slightly hairy on both surfaces, whitened beneath. White flowers in flat-topped terminal clusters bloom from May to July. The round berries are *white*, about ¼ inch in diameter. They ripen from July to September.

Habitat: This dogwood grows in damp areas from Alaska to Greenland, south to New Jersey and Minnesota and, in the western mountains, to New Mexico and California.

Remarks: The twigs of this shrub are eaten by deer, elks, moose, cottontails and snowshoe hares.

Red-Osier Dogwood

Berry use: The berries can be eaten cooked, but caution is urged in consumption of raw berries. Songbirds, game birds, black bears and other mammals eat them.

Similar species: GRAY-STEMMED DOGWOOD (*C. racemosa*), ROUGHLEAF DOGWOOD (*C. drummondii*) and, occasionally, ROUNDLEAF DOGWOOD (*C. rugosa*) also have *white* berries. See also Dogwood, Blue Berries.

WOODY VINES AND SHRUBS • HEATH FAMILY

CREEPING SNOWBERRY
Gaultheria hispidula

Other names: SNOWBERRY WINTERGREEN

Description: This is a creeping, mat-forming shrub with slender, brown, hairy stems. The small, short-stalked leaves are alternate, pointed, with rolled edges and a wintergreen odor when crushed. They are untoothed, leathery above, brown-hairy underneath. Small, white, solitary flowers bloom in the leaf axils. The berries, which ripen in August or September, are round and *white*, usually with some bristly brown hairs present. They, too, have a wintergreen odor when crushed.

Habitat: Creeping Snowberry grows in cold, mossy woods and bogs from Newfoundland to British Columbia, south to New England and West Virginia, Minnesota and Idaho.

Berry use: The berries taste like wintergreen and may be used raw or cooked. They are also enjoyed by thrushes and grouse.

DEERBERRY
Vaccinium stamineum

Other names: TALL DEERBERRY, SQUAW-HUCKLEBERRY

Description: This deciduous shrub, 2 to 10 feet tall, has downy, reddish purple branchlets. The leaves are toothless, thin, elliptic or egg-shaped, whitish underneath. Greenish white, bell-shaped flowers bloom in drooping clusters with leaflike bracts. The fruits are *greenish or purple*, sometimes whitened with a bloom. They ripen from July to September.

Habitat: Look for this shrub in sandy or rocky woods and thickets from Massachusettes to Ontario, southwest to Kansas, and south to Texas and Florida.

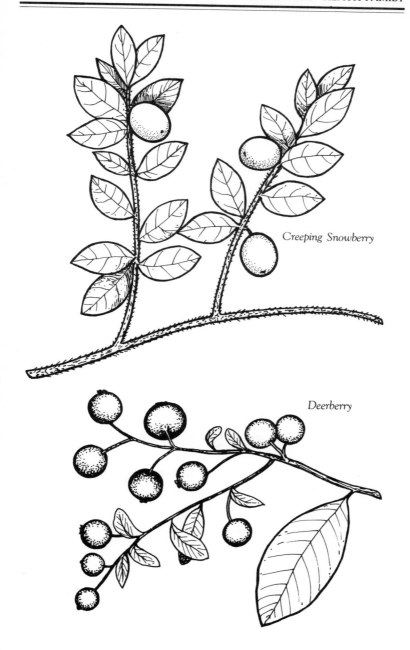

Creeping Snowberry

Deerberry

Remarks: Deerberries can be distinguished from blueberries by the fact that deerberries lack the warty speckles on the twigs that characterize blueberries. They can be told from huckleberries by the lack of resin-dotted foliage that characterizes huckleberries.

Berry use: Although the berries are sour and considered inedible raw, they are edible when cooked.

Similar species: LOW DEERBERRY (*V. caesium*) does not grow as high. It is found in dry soils from the mountains of Pennsylvania and West Virginia to Florida and Lousiana.

WOODY VINES AND SHRUBS • HONEYSUCKLE FAMILY

SNOWBERRY
Symphoricarpos albus

Other names: WAXBERRY, THIN-LEAVED SNOWBERRY

Description: This is a finely branched deciduous shrub about 1 to 5 feet tall. It has opposite, short-stalked leaves that are elliptic to roundish, bluntly pointed at both ends. Inconspicuous bell-shaped, pink flowers bloom in the leaf axils or terminal clusters. The round, waxy, *white* berries, ¼ to ½ inch in diameter, grow in clusters. They ripen from August to October.

Habitat: This shrub grows on dry, rocky soil from Quebec to British Columbia, south to Virginia, Nebraska, and California.

Remarks: Snowberry is often cultivated for the sake of its attractive berries.

Berry use: Be cautious about tasting these berries. It is said that even birds do not care for their peculiar flavor.

Similar species: WOLFBERRY (*S. occidentalis*), also called BUCKBRUSH or WESTERN SNOWBERRY, is similar but has larger leaves and the flowers have no stalks. The berries are dull, *greenish white*, but they soon *become discolored and blackish.* The berries often remain on the shrub throughout the winter. This is one of the few woody plants that grows freely on open prairies. The range is from New England to British Columbia, south to northern Georgia and New Mexico.

Snowberry

TREES

WHITE MULBERRY
(Rose Family) See Purple Berries.

TREES • LAUREL FAMILY

CALIFORNIA LAUREL
Umbellularia californica

Other names: OREGON MYRTLE, CALIFORNIA BAY

Description: This is a small to medium-sized tree, easily recognized by its smell, which has been described as similar to that of bay rum. It is a beautiful tree that grows 20 to 80 feet tall with a broad, rounded, dense crown. Its leaves are alternate, evergreen, lance-shaped or elliptic, thick and leathery. Yellow-green flowers bloom in clusters, appearing with the leaves. Bees are much attracted by them. The berrylike fruit *is greenish to purple,* about 1 inch long.

Habitat: California Laurel grows best on moist bottom lands in Oregon and California.

Use of berrylike fruit: The berries can be eaten raw or roasted.

California Laurel

Dark purple and purplish black are hard to distinguish. Purple sometimes verges on blue or red. Check other sections if necessary.

FLOWERING HERBS • LILY FAMILY

NODDING WAKEROBIN
Trillium cernuum

Other names: NODDING TRILLIUM

Description: Like other trilliums (see Red Berries), Nodding Wakerobin has leaves in a single whorl of 3 and a large, solitary flower. The flower stalk bends downward, so that the flower, which is 1 to 1¼ inches broad, nods beneath the leaves. The white or pale-pink petals recurve at the tip and have wavy margins. The berry is oval and *reddish purple.*

Habitat: Rich woodlands are home for this species, which grows from Newfoundland to Ontario and Manitoba, south to Georgia and Missouri.

Berry use: The berries are inedible.

INDIAN CUCUMBER-ROOT

See Blue Berries.

POKEWEED

Phytolacca americana

Other names: POKE, SCOOT, PIGEON BERRY, INKBERRY GARGET, POCAN, POKEBERRY

Description: The smooth, reddish or purplish stems of this strong-smelling weed are a distinguishing characteristic. It is widely branched and grows from 3 to 10 feet tall, with large, smooth, alternate leaves. Opposite the leaves grow long-stalked, erect, terminal clusters of flowers with 5 white, petal-like sepals. The plant is notable for the clusters of *dark purple, purplish red or purplish black* berries that persist well into winter. The berry clusters, 4 to 7 inches long, droop from red stalks with their weight.

Habitat: Pokeweed flourishes in damp thickets, clearings, cultivated fields, roadsides and waste places. Its range is from Minnesota to Quebec, south to Florida and Texas.

Remarks: Indians used the powdered roots as treatment for fever and various illnesses. Early settlers made poultices of the dried leaves and used them on wounds, swellings and skin disturbances. The young shoots are occasionally used as a vegetable, cooked with two or more changes of water. Roots, seeds, mature stems and leaves contain phytolaccin and are dangerously *poisonous.* They have a laxative and narcotic effect and can cause death.

Berry use: The berries are *poisonous,* especially when green. If cooked, ripe berries can reportedly be used, as in pies. There is less poison in the berries than in other parts of the plant. Pioneers used the berry juice for treatment of skin eruptions and skin ulcers. The juice, which is crimson, was also used as a dye and to make ink, thus giving the plant one of its popular names. Birds are fond of the berries. They were a favorite food of the now extinct passenger pigeon, hence the popular name PIGEON BERRY.

Pokeweed

PRICKLY PEAR

(Cactus Family) See Red Berries.

FLOWERING HERBS • GINSENG FAMILY

AMERICAN SPIKENARD

Aralia racemosa

Other names: LIFE-OF-MAN, SPIKENARD, OLD MAN'S ROOT

Description: This is a spineless, branching plant with a smooth, blackish stem. It grows 3 to 6 feet tall. The leaves are large and spreading, with many leaflets. The leaflets are oval or heart-shaped, pointed at the tip. Small, white flowers bloom in round umbels along a central stem. They are followed by small, dull, *brownish purple or brownish crimson* berries which mature in August or September.

Habitat: Spikenard grows in rich, moderately acid soils in woods and thickets from Quebec to Manitoba, south to Kansas, Missouri and Georgia.

Remarks: The roots of this plant are stout, fleshy and aromatic. They were once used for medicinal purposes and to flavor root beer.

Berry use: The berries can be used for jelly but are *poisonous if raw.*

Similar species: CALIFORNIA SPIKENARD (*A. californica*) is a larger plant with fleshy stems and huge, pinnately compound leaves (1 to 3 feet long). It bears its round flower umbels in branching racemes at the top of the plant. It is found in California and southern Oregon. See also Black Berries for Wild and Bristly Sarsaparilla.

SMOOTH GROUND CHERRY

(Nightshade Family) See Red Berries.

Spikenard

WOODY VINES AND SHRUBS

COMMON JUNIPER
(Cedar or Cypress Family) See Blue Berries.

COMMON BARBERRY
(Barberry Family) See Red Berries.

OREGON GRAPE
(Barberry Family) See Blue Berries.

WOODY VINES AND SHRUBS • SAXIFRAGE FAMILY

SWAMP BLACK CURRANT
Ribes lacustre

Other names: SWAMP CURRANT, SWAMP GOOSEBERRY, BRISTLY CURRANT, BRISTLY BLACK CURRANT

Description: This species is 1 to 3 feet tall, with prickles and weak thorns. The stems have a foul odor when broken. The leaves are 3- to 5-lobed, toothed, nearly hairless on both surfaces. Green or purplish flowers hang in a drooping cluster. The fruits are *purple or purplish black,* bristly and foul-smelling when crushed. See Red Berries for a general description of currants.

Habitat: Swamp Black Currant grows in moist, often rocky woods and bogs from Newfoundland to Alaska, south to Massachusetts, the Great Lakes region, the Black Hills, Colorado and California.

Berry use: The berries are enjoyed by nearly all birds and some animals, but humans find them scarcely edible.

GOOSEBERRIES
Ribes spp.

Gooseberries and currants both have 3- to 5-lobed maplelike, coarse, alternate leaves. Gooseberries differ from currants in that they have 1 to 3 thorns at the bases of the leafstalks. Gooseberry flowers are borne in shorter clusters than those of currants. The round, many-seeded berries of the gooseberry are bristly and firmly attached to their stalks, while

Swamp Black Currant

those of the currant are usually not bristly and are easily detached from the stalks. Gooseberries are rich in pectin and all species are edible. Here are some gooseberries with purple berries:

ROUNDLEAF GOOSEBERRY
Ribes rotundifolium

Other names: MOUNTAIN GOOSEBERRY

Description: This is a short-thorned shrub 2 to 5 feet tall with round-lobed leaves, smooth slightly hairy underneath. Greenish purple flowers develop into smooth, *purplish or black* berries that ripen from June to September.

Habitat: This species is found in cool, rocky mountain woods from Massachusetts to North Carolina and eastern Tennessee.

Berry use: The berries may be used for pies, sauce or jelly. Berries that are slightly underripe are richer in pectin and best for jelly or jam.

PRICKLY GOOSEBERRY
Ribes cynosbati

Other names: PASTURE GOOSEBERRY, DOGBERRY

Description: This shrub grows to 4 feet tall. The leaves are similar to the preceding species, except that they are softly hairy on both sides. There usually are no prickles and the thorns are short. Greenish flowers give way to *reddish purple* berries covered with long spines. They ripen from July to September.

Habitat: Look for this gooseberry in rocky woods and clearings from New Brunswick to Manitoba, south to Georgia, Alabama and Missouri.

Berry use: These berries are sweet and tasty, but they should be cooked to soften the long spines.

Similar species: MISSOURI GOOSEBERRY (*R. missouriense*) is much like the preceding two, except that it grows to 6 feet. Its stout thorns are often red. The berry is *purplish or black* and smooth. This species does not grow quite as far south as Prickly Gooseberry. NORTHERN GOOSEBERRY (*R. oxycanthoides*), also called CANADA GOOSEBERRY, is a low-growing shrub with berries that are round and *reddish purple*, about ⅜ inch in

Roundleaf Gooseberry

Prickly Gooseberry

Missouri Gooseberry

Northern Gooseberry

diameter. When they ripen in July or August, they are sweet. This is a northern species, growing mostly in coniferous woods from Canada south to Michigan, Minnesota, South Dakota and Montana. SIERRA GOOSEBERRY (*R. roezlii*) is a thorny shrub with berries about ½ inch in diameter. They are covered with stiff spines and are *reddish purple* when mature, from July to September. They grow in forests and logged-over areas of the Sierras, and make excellent jelly. See also Black Berries for Smooth Gooseberry.

WOODY VINES AND SHRUBS • ROSE FAMILY

PURPLE CHOKEBERRY
Pyrus floribunda

This shrub is similar to Black Chokeberry except that its fruits are *reddish purple to purplish black*. They are edible and can be used like blueberries. But do not mistake them for buckthorn berries! See Black Chokeberry in Black Berries.

JUNEBERRIES
Amelanchier spp.

Other names: SERVICEBERRY, SARVISBERRY, SHADBUSH, SHAD-BERRY, SASKATOON BERRY, INDIAN PEAR, SUGAR PEAR, SUGARPLUM

Description: These are deciduous shrubs or small trees. Their leaves are simple, alternate, toothed, usually blunt-tipped. Their buds are unique in being pink or reddish, slender, dark-tipped and usually somewhat twisted. Fragrant, showy flowers with 5 long petals nod in clusters. They often appear before the leaves. The berrylike pomes, which ripen from June to September, are pear-shaped or round and can sometimes be found on the bushes in late winter. They resemble blueberries but are darker, larger and plumper. *Purple is the most typical color, but they range from red or blue to purplish black.*

Habitat: Juneberries are found in woods, thickets, prairies, along streams and roads and in swamps from Newfoundland to Alaska and south to California and Georgia.

Remarks: Although juneberries are not difficult to recognize as a group,

Juneberry

the individual species are often difficult to identify. There are about 25 species. The name Shadbush occurs because in some parts of the country the plant blossoms about the time the shad begin their migration upstream to spawn.

Use of berrylike fruit: Juneberries are one of the first wild fruits of the season and they are not used as much as they could be. Fresh or dried, they were an important food for the Indians, who used them in pemmican. They are sweet, rather bland and seedy. They can be used much the same way as blueberries. Pectin should be added for jam or jelly. During cooking, the 10 seeds soften and give the fruit an almondlike flavor. The fruits are a favorite food of bears, grouse, pheasants, robins and other songbirds.

BLACK RASPBERRIES and BLACKBERRIES
See Black Berries.

PROSTRATE SAND CHERRY
Prunus depressa

Other names: SAND CHERRY, DWARF CHERRY

Description: For a general description of cherries and plums, see Red Berries. This is a prostrate, spreading shrub with branchlets that rise 8 inches to 2 feet high. Its leaves are narrowly elliptic, toothed above the middle, light green above, paler underneath. The white flowers grow 2 to 4 in a cluster. The berrylike fruits are pea-sized, *reddish purple or blackish purple.* They ripen from July to September.

Habitat: Sand Cherries grow in gravelly or sandy soil from Quebec to Ontario, south to Pennsylvania and west to Wisconsin.

Use of berrylike fruit: The fruit is bitter but edible if the seeds are discarded.

BEACH PLUM
Prunus maritima

Description: This straggling, thornless shrub, 1 to 8 feet tall, has velvety branchlets that become smooth with time. The egg-shaped leaves are sharply toothed, smooth on top, downy below. Lovely white flowers bloom in early spring, before the leaves are out. The fruit is ½ to 1 inch

Prostrate Sand Cherry

in diameter, *purple* with a whitish bloom, ripening from August to October.

Habitat: This plum tree inhabits coastal plains from Maine south to Delaware and along the ridges of the Appalachians. It is especially common on Cape Cod.

Use of berrylike fruit: The plums are sweet and juicy and may be used in preserves, jellies and sauce.

FLATWOODS PLUM
Prunus umbellata

Other names: HOG PLUM, BLACK SLOE

Description: This shrub or small tree has spiny-tipped branches. The leaves are elliptic, pointed at the tip, finely toothed. White flowers, 2 to 4 in a cluster, are followed by fruits, round, *reddish purple*, ½ inch or less in diameter. They ripen in June or July.

Habitat: Flatwoods Plum thrives in dry, sandy woods or along stream banks from North Carolina to Florida and Texas, north in the Mississippi Valley to Arkansas.

Use of berrylike fruit: Although bitter, the fruits can be used in jellies and sauce. See Red Berries and Black Berries for other cherries and plums.

BLACK CROWBERRY
(Crowberry Family) See Black Berries.

PRICKLY PEAR and PINCUSHION CACTUS
(Cactus Family) See Red Berries.

BURNING-BUSH
(Bittersweet Family) See Red Berries.

Beach Plum

WOODY VINES AND SHRUBS • VINE FAMILY

PEPPER VINE
Ampelopsis arborea

Description: Pepper Vine can be either a bushy, somewhat upright shrub or a high-climbing vine. It is one of the few plants that have leaves which are divided into leaflets that are again divided into smaller leaflets. These leaflets are coarsely toothed, more or less egg-shaped. Opposite some of the leaves are tendrils. Small, greenish flowers are borne in long-stalked, forking clusters. The berries are round, *dark purple to black*, about ¼ inch in diameter.

Habitat: Rich, moist woods and thickets are home to the Pepper Vine, from Maryland to Missouri, south to Texas and Florida.

Berry use: The berries are bitter and inedible.

WILD GRAPES
Vitis spp.

Grapes are thornless, high-climbing vines with shredding bark, alternate leaves and forked tendrils. The branchlets usually have a brown pith, which is interrupted at the leaf nodes by partitions. The leaves are large, prominently veined, coarsely toothed, heart-shaped at the base and often lobed. The flowers are green or greenish yellow, small and fragrant. They are borne in compact clusters opposite the leaves, appearing in the spring as the leaves begin to expand. The fruits are round, juicy, few-seeded berries. They are *purple, blue, black or amber*. They are an important food for wildlife. Man also enjoys wild grapes. They are too tart to enjoy eating raw, but they make excellent juice, jam and jelly. They have plenty of pectin if picked before fully ripe.

Caution! Before you decide a vine is wild grape, check the seeds. Grapes have several small seeds. A similar plant, Common Moonseed, has one flat, crescent-shaped seed. Moonseed berries are poisonous.

There are numerous species of grapes. Listed below is the species that has the largest berries and is considered the most desirable. It is the ancestor of Concord and other cultivated grapes.

Pepper Vine

FOX GRAPE
Vitis labrusca

Description: This species, from which the Concord Grape was derived, has a tendril or flower cluster opposite each leaf. (Most grapes have nothing opposite every third leaf.) It is a high-climbing vine with lobed, heart-shaped, somewhat leathery leaves that are dark green above and coarsely but shallowly toothed. The leaf undersides and twigs are rusty-woolly. The berries are ½ to 1 inch in diameter, *purple, black or rarely, amber.* They mature from August to October.

Habitat: Fox Grape grows in woods, thickets and along streams from Maine to Michigan south to Florida and Mississippi.

Remarks: Grapevines can be parasitic because they cover the foliage of the shrubs or trees upon which they climb, depriving them of sunlight. The supporting plants are sometimes killed, and the grapes may form dense, impenetrable tangles that stretch for acres.

Berry use: Wild grapes can be used in the same way as cultivated varieties except that they usually require more sweetening. They have plenty of pectin before fully mature, so when making grape jelly or jam, have ¼ of the grapes underripe.

Similar species: CALIFORNIA WILD GRAPE (*V. californica*) is one of the western species; it has purple grapes that are large and sweet. It grows in southwest Oregon and most of California.

WOODY VINES AND SHRUBS • HEATH FAMILY

SALAL
Gaultheria shallon

Other names: SHALLON BERRY

Description: Salal is a shrub with sticky, hairy stems that grows 1 to 10 feet tall. It sometimes forms dense thickets in evergreen forests. The leaves are alternate, oval, leathery and thick. Pinkish, bell-shaped flowers hang near the end of the stems from May to July. The berries that appear in August stay on the shrub until October. They are round, hairy, *dark purple or black.*

Habitat: This shrub grows along the Pacific Coast from British Columbia

Fox Grape

to California and east to the western slopes of the Cascades. It is frequent in the redwood forests of California.

Remarks: Salal is a western species of wintergreen, with berries that are larger and less spicy than its eastern relative. "Salal" is a corrupted form of its Indian name.

Berry use: The berries of this plant played an important part in the diet of the Indians who lived in the area where they are found. They boiled them down for sugar, dried them or made them into cakes. The journal of the Lewis and Clark Expedition contains several references to their being eaten by the Indians of what is now Oregon.

At present, Salal berries are used for jelly, syrup and survival food. Although the skin is thick, the berries have a good flavor.

ALPINE BEARBERRY
Arctostaphylos alpina

This mat-forming shrub resembles Bearberry (see Red Berries) except that its leaves are wavy-toothed and thinner. They wither but usually remain attached in winter. The fruit is barely edible. This species is found growing on poor soil from the Arctic to Newfoundland and in high altitudes of Quebec, Maine and New Hampshire. Its berries are *purplish or purplish black.*

THINLEAF BILBERRY
Vaccinium membranaceum

This shrub, also called SQUARE-TWIG BILBERRY, BIG WHORTLE-BERRY and MOUNTAIN HUCKLEBERRY, has berries that are *reddish purple or black* and are rather sour but edible. It grows from Ontario to British Columbia, south to California and east to the Black Hills of South Dakota. See Bilberries, Blue Berries.

BLUEBERRIES, HUCKLEBERRIES
See Blue Berries.

DEERBERRIES
See White and Green Berries.

Salal

WOODY VINES AND SHRUBS • VERVAIN FAMILY

BEAUTYBERRY
Callicarpa americana

Other names: JEWELBERRY, FRENCH MULBERRY

Description: This shrub is often grown as an ornamental for the beauty of its berries. It is deciduous, about 3 to 6 feet tall, with ashy gray branchlets that are hairy or scaly. The leaves are opposite, oval or egg-shaped, tapering to a point at both ends, sharply toothed. The small, funnel-shaped flowers are bluish. Both flowers and berrylike fruits grow in dense axillary clusters. The fruits are *bright violet or magenta* and they mature from August to October.

Habitat: Sandy or rocky soil, especially where it is moist, is favored by Beautyberry. The range is from Maryland to Oklahoma, south to Texas and Florida.

Use of berrylike fruit: The fruits are juicy and can be eaten raw, but they are not good.

HONEYSUCKLE, CORALBERRY
(Honeysuckle Family) See Red Berries, Blue Berries, and Black Berries.

BLACKHAW, VIBURNUMS, COMMON ELDERBERRY
(Honeysuckle Family) See Black Berries.

TREES • ELM FAMILY

COMMON HACKBERRY
Celtis occidentalis

Other names: AMERICAN HACKBERRY, NETTLE TREE, HOOP-ASH

Description: Hackberries are often mistaken for their close cousins, the elms. One way to recognize a hackberry is by the witches'-brooms, caused by a fungus disease, that frequently exist high in the tree. Common Hackberry is a small to large tree with corky lumps on its gray to brown bark. Its leaves are alternate, long-pointed, coarse-toothed, often lopsided and usually rough and hairy above. Inconspicuous green flowers are succeeded by oval, pea-sized, berrylike drupes. *Orange at first,* they turn *deep purple or dark red* at maturity.

Beautyberry

Habitat: This species grows in woods and open places from Quebec to Manitoba, south to Georgia and Texas.

Remarks: Common Hackberry is often planted in the West because of its resistance to drought. Although it grows best on rich, moist soils, it adapts itself well to poor, dry sites.

Use of berrylike fruit: The fruits are edible raw, dried or cooked. The pulp is thin, tasty and sweet. Cedar waxwings are particularly fond of the fruits. They are also enjoyed by other birds and such animals as squirrels, beavers, raccoons and opossums.

Similar species: There are about 70 species of hackberry shrubs or trees. See Red Berries for Sugarberry.

TREES • MULBERRY FAMILY

RED MULBERRY
Morus rubra

Other names: VIRGINIA MULBERRY, BULBERRY

Description: This is a small, deciduous tree, usually 20 to 40 feet high, with a dense, round-topped crown. Its alternate leaves are broadly oval, coarsely toothed, often with 2 or 3 lobes. They may be somewhat like sandpaper on top, hairy below. Drooping spikes of small, green flowers appear with the leaves. Male and female flowers may both be present on the same tree. The fruit is 1 to 1¼ inches long and resembles an elongated blackberry. It is *red at first, purple when ripe,* in June or July.

Habitat: Red Mulberry trees like rich, moist soil. They range from Massachusetts to South Dakota, south to Florida and Texas.

Use of berrylike fruit: Unripe fruit contains hallucinogens, but ripe fruit can be eaten raw, though it is tart. It is more often used for jelly, with pectin added, or pie. Birds often pluck all the berries before humans harvest them. The fruits attract many of our favorite songbirds.

Similar species: WHITE MULBERRY (*M. alba*) was introduced to this country before Revolutionary days in a futile attempt to start a silk industry in the New World. It is now common in the East. Its berrylike fruits are *white, pink or purple.* They are sweet and edible, but some people think raw berries are insipid and prefer them dried and added to hot breads

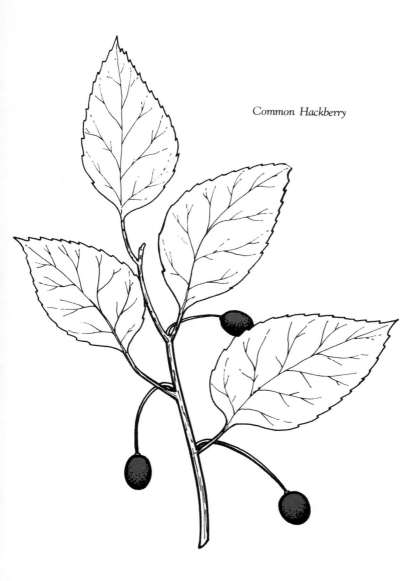

Common Hackberry

and muffins. PAPER MULBERRY (*Broussonetia papyrifera*) has fleshy red fruit that is barely edible. It is an ornamental tree but also grows wild.

FLORIDA STRANGLER FIG
See Yellow or Orange Berries.

CALIFORNIA LAUREL
(*Laurel Family*) See White or Green Berries.

OREGON CRAB APPLE
(*Rose Family*) See Yellow or Orange Berries.

HAWTHORN
(*Rose Family*) See Red Berries.

TREES • SOAPBERRY FAMILY

BUTTERBOUGH
Exothea paniculata

Description: This tree has reddish brown, scaly bark and grows to 50 feet tall. Its leaves are pinnately compound with 2 to 6 wavy-margined leaflets 4 to 6 inches long. They are evergreen and are arranged alternately on the branch. The flowers are white, the berries *purple, occasionally orange,* and fleshy.

Habitat: This is a tree of eastern and southern Florida.

Berry use: It is best to leave these berries alone.

TREES • TUPELO FAMILY

WATER TUPELO
Nyssa aquatica

Other names: TUPELO GUM, COTTON GUM

Description: This tree is similar to Black Tupelo (see Blue Berries) except that it grows taller (100 feet or more) and has larger leaves and its fruits are larger (about 1 inch long), oval-shaped rather than round. The fruits are *purple* and have conspicuously ribbed pits.

Red Mulberry

Habitat: This is a tree found in swamps and bottomlands, often standing in water, in coastal plains from Virginia to northern Florida, west to Texas and northward in the Mississippi Valley to southern Illinois.

Use of berrylike fruit: The fruits are edible raw but are bitterly pungent. They are sometimes used for making preserves.

COMMON PERSIMMON

(Ebony Family) See Yellow or Orange Berries.

TREES • OLIVE FAMILY

FRINGETREE

Chionanthus virginicus

Other names: OLD MAN'S BEARD, GRANDFATHER GRAYBEARD, GRAYBEARD TREE

Description: This is a small tree or shrub that grows 30 feet tall. It is a handsome tree with large leaves, opposite one another, that turn yellow in the fall. The leaf margins are smooth or wavy; underneath, the veins are hairy. The tree is often planted as an ornamental because of the lovely, fragrant flowers that appear in May or June. The 4 to 6 white petals are so long and narrow that they appear fringelike, giving the tree its name. They are borne in drooping clusters. The berrylike fruits, oval and about 1 inch long, are *dark purple or dark blue*, often with a white bloom. They mature from September to October.

Habitat: This tree grows in woods and on floodplains, and is native from New Jersey to Florida, west to Texas and north to Oklahoma, Missouri and Ohio.

Use of berrylike fruit: None known.

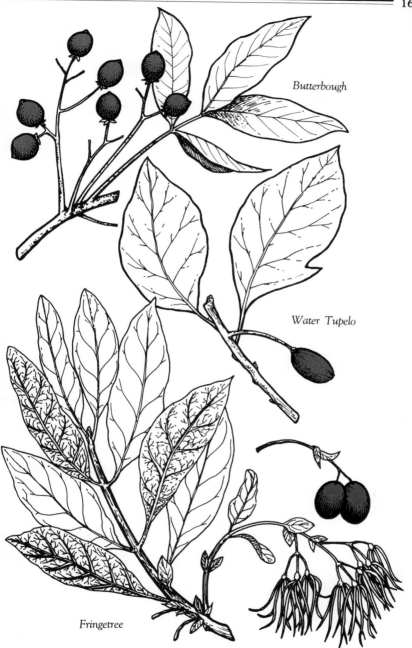

Butterbough

Water Tupelo

Fringetree

Black berries may go through red, blue or purple stages. Check other sections if necessary.

FLOWERING HERBS • LILY FAMILY

WHITE CLINTONIA
Clintonia umbellulata

This plant is similar to Clintonia except that is has smaller flowers, white, dotted with green and purple. The berries are *black* and are not considered edible. White Clintonia grows in rich woods from New York to Ohio, and south in the mountains to Georgia. See Clintonia, Blue Berries.

STARRY FALSE SOLOMON'S-SEAL
See Red Berries.

SOLOMON'S-SEAL

See Blue Berries.

LARGE-FLOWERED TRILLIUM

Trillium grandiflorum

Other names: WHITE TRILLIUM, LARGE-FLOWERED WAKEROBIN

Description: For a general description of trilliums, see Red Berries. This is the most striking of the trilliums because of the size of its solitary flower — 2 to 3 inches broad. The white petals, which are much longer than the green sepals, turn pink with age. The berry is round, *black*, slightly 6-lobed, ¾ to 1 inch in diameter.

Habitat: Large-Flowered Trillium is found in rich woods, especially ravines and wooded upland slopes, from Quebec to Ontario and Minnesota, south to North Carolina and Missouri.

Berry use: The berries are inedible.

CARRION-FLOWER

Smilax herbacea

Description: This herbaceous vine climbs by means of tendrils to 7 feet or more. It is similar to the greenbriers but lacks thorns and is not woody. Its leaves are heart-shaped with parallel veins. The most distinctive trait of this vine is that its flowers smell putrid. This is probably an evolutionary adaptation to attract flies for purposes of fertilization. The flowers are greenish yellow, clustered at the end of a long stem, as are the berries. The berries are *black or dark blue;* they ripen from August to October.

Habitat: Carrion-Flower grows in moist to dry woods and fence rows from southern Canada to Georgia, Missouri and Colorado.

Remarks: The spring shoots can be prepared and eaten like asparagus. The edible rootstocks are used as food throughout the year.

Berry use: The berries can be eaten raw or cooked into pie, jelly or sauce. No pectin is needed for jelly.

Carrion-Flower

FLOWERING HERBS • IRIS FAMILY

BLACKBERRY LILY
Belamcandra chinensis

Description: This is a 6-petaled, lilylike flower with leaves like that of the iris family, to which it belongs. It branches out more than the typical iris, however. The flowers are smaller and flatter than those of a lily. They are light golden-orange, mottled with dull magenta spots. The fruit capsule holds its *black* seeds in an upright, thimble-shaped mass which resembles a blackberry.

Habitat: This plant has escaped from cultivation to roadsides and thickets from Connecticut and Indiana south to Georgia and Missouri.

Use of berrylike fruit: It is not advisable to eat these "berries."

POKEWEED
(Pokeweed Family) See Purple Berries.

JUNEBERRIES
(Rose Family) See Purple Berries.

FLOWERING HERBS • GINSENG FAMILY

WILD SARSAPARILLA
Aralia nudicaulis

Description: This leafy plant grows about 1 foot tall. A single compound leaf is borne on a long stalk that rises near the root. It divides into 3 leaflets, which divide again into 3 or 5 leaflets that are egg-shaped and toothed. Small, greenish white flowers bloom on separate, shorter leafless stems. They form 3 terminal, rounded clusters. In late summer, the berries turn *purplish black.*

Habitat: This is a common perennial of forests. It grows across Canada south to Georgia and Colorado.

Remarks: The horizontal roots are aromatic and are sometimes used as a substitute for true Sarsaparilla, a tropical plant, as in root beer.

Berry use: The berries are *poisonous to humans if eaten raw* but can be used for jelly. Bears and foxes enjoy them.

Blackberry Lily

Common Nightshade

Wild Sarsaparilla

Similar species: BRISTLY SARSAPARILLA (*A. hispida*), also called WILD EL-
DER, is woody at the base stem. Its berries are *purplish black, poisonous
when raw* but all right for jelly. It grows across Canada south to West
Virginia, Indiana and Minnesota.

FLOWERING HERBS • NIGHTSHADE FAMILY

COMMON NIGHTSHADE
Solanum nigrum

Other names: BLACK, DEADLY or GARDEN NIGHTSHADE, POISON-
BERRY

Description: This is a bushy weed, growing to 2 feet tall. The leaves are
roughly triangular, deep green, long-stalked and irregularly toothed.
The small, white flowers grow in drooping clusters. The 5 petals turn
back and the stamens form a protruding yellow beak. Clusters of round,
green berries turn *black* when mature.

Habitat: This plant is common over most of the United States and
Canada, especially where the ground has been disturbed.

Remarks: Atropine, a drug used to dilate pupils of the eyes, can be
extracted from Common Nightshade, as well as from several related
species.

Berry use: Although there are reports of people cooking the fully ripe
berries in pies, this is not advisable, since they are *poisonous*. The green
berries contain highly dangerous amounts of the poison solanine. Even
three berries can be fatal to a child. See also Bitter Nightshade, Red
Berries.

WOODY VINES AND SHRUBS

COMMON JUNIPER
(Cedar or Cypress Family) See Blue Berries.

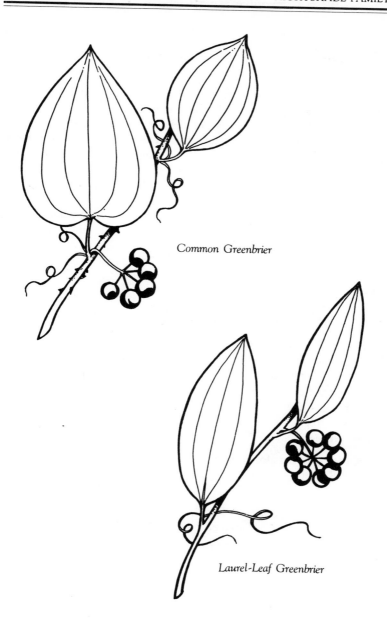

Common Greenbrier

Laurel-Leaf Greenbrier

WOODY VINES AND SHRUBS • LILY FAMILY

GREENBRIERS
Smilax spp.

Greenbriers are green-stemmed, often prickly, woody vines. They climb by means of paired tendrils that rise from the base of the leafstalks and remain after the leaves fall. The leaves are alternate, with prominent parallel veins. On some southern species, they remain on through the winter. Small, yellowish green flowers are borne in clusters at the end of stalks that rise from the leaf axils. The berries are small, *black, bluish or red.* Songbirds and game birds eat the berries. Greenbriers are regarded as one of the most valuable of all shrubs and vines for wildlife food and cover. Although they are often considered inedible by humans, the berries can be eaten raw or cooked. Here are some greenbriers with black berries:

COMMON GREENBRIER
Smilax rotundifolia

Other names: CATBRIER, HORSEBRIER, ROUND-LEAF BRIER

Description: This scrambling vine is armed with stout, broad-based prickles, especially on its lower branches. The stems and branchlets are either round or 4-angled. The leaves are egg-shaped or nearly round, often with a heart-shaped base, green and glossy on both sides. Small, yellowish green flowers bloom in the leaf axils. By September, they have been replaced by *blue-black* berries, about ¼ inch in diameter, with a powdery bloom.

Habitat: Greenbriers form almost impenetrable thickets, reminding one of Peter Rabbit's "brier patch." Common Greenbrier ranges from Nova Scotia to Minnesota, south to Florida and Texas.

Remarks: The young shoots and leaves of greenbriers may be cooked and eaten like asparagus or used in salads.

Berry use: The berries are edible raw but are better when cooked.

Similar species: BRISTLY GREENBRIER (*S. tamnoides* var. *hispida*) is a high-climbing species with round stems and weak bristles. The berries are *black.* GLAUCOUS GREENBRIER (*S. glauca*) has round, prickly stems

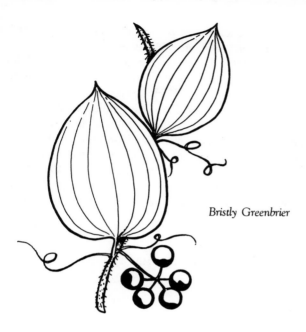

Bristly Greenbrier

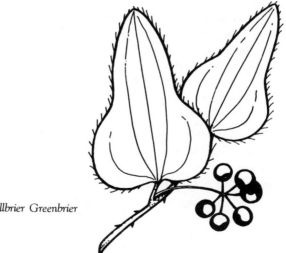

Bullbrier Greenbrier

whitened with a bloom. The leaves are whitened underneath. The berries are *blue-black or blue*, with a bloom. BULLBRIER *(S. bona-nox)*, also called CHINABRIER or SAWBRIER, has leaves that are triangular to fiddle-shaped, green beneath but often mottled with white. The 4-angled stems have many stout prickles. The *black* berries often persist until spring. LAUREL GREENBRIER *(S. laurifolia)*, often called BAMBOO-VINE or, because of its fierce thorns, BLASPHEME-VINE, has leathery, evergreen leaves. The stems are prickly near the base and on the more vigorous shoots. The *black* berries do not ripen until the second year, so green and black berries may be present at the same time. WILD BAMBOO *(S. auriculata)* is evergreen and has 4-angled stems and zigzag branches. The leaves are usually fiddle-shaped. The berries are *black* with a bloom.

WOODY VINES AND SHRUBS • MOONSEED FAMILY

COMMON MOONSEED

Menispermum canadense

Other names: CANADA MOONSEED

Description: This thornless, deciduous vine climbs by green, faintly grooved, twining stems. It is nearly hairless and has shallowly lobed, toothless leaves that are dark green and smooth above, downy underneath. Small, greenish white flowers bloom in June and July. The fruits, which ripen in September, are pea-sized, berrylike drupes, *bluish black or dark blue*, with a bloom. They hang in loose clusters and resemble grapes but can be distinguished from grapes by the fact that each fruit has single, flattened, crescent-shaped seed.

Habitat: This shrub grows along streams and fence rows and in woods from New England and Quebec to Manitoba, south to Georgia and Oklahoma.

Remarks: Reports of poisoning by "wild grapes" are probably a result of eating Common Moonseed fruits, which are poisonous. Moonseed resembles members of the grape family in the size, shape and color of the fruit, the leaves and the fact that it is also a vine. As noted, it can be distinguished by its seed and also by the fact that it climbs by twining stems rather than tendrils.

Common Moonseed

Use of berrylike fruit: This is a *poisonous* fruit and should not be eaten. Fortunately, this vine is rather rare.

Similar species: CUPSEED (*Calycocarpum lyoni*) is a rather smooth, high-climbing southern vine. The leaves are large and have 3 to 5 deeply cut lobes. The fruit ripens from August to September, is roundish to oval and *black*. It is nearly an inch long and has a single, flattened, dish-shaped seed. *Do not eat the fruit.*

OREGON GRAPE
(Barberry Family) See Blue Berries.

WOODY VINES AND SHRUBS • SAXIFRAGE FAMILY

WILD BLACK CURRANT
Ribes americanum

Other names: AMERICAN BLACK CURRANT

Description: See Red Berries for general information on currants. This upright shrub is characterized by yellow resin dots on both leaf surfaces as well as on twigs, buds and fruits. It is thornless and grows to 4 or 5 feet. The leaves have 3 to 5 lobes and are double-toothed. Large yellow or white flowers bloom in drooping clusters. The berries, which ripen from June to September, are *black*, smooth and resin-dotted, pea-sized.

Habitat: This species grows in open woods and flood plains fron Nova Scotia to Alberta, south to West Virginia, the Great Lakes area, Oklahoma and New Mexico.

Berry use: The berries are used for jelly, jam and wine.

Similar species: GOLDEN CURRANT (*R. odoratum*), also called BUFFALO CURRANT or CLOVE BUSH, is a tall, upright, thornless currant with branchlets covered with grayish down. The name comes from the golden-yellow flowers, which have a delightful spicy fragrance. The berries are smooth and *black, occasionally yellow*. They ripen from June to August. This species grows east of the Rockies. A species with the same popular name but a different scientific name, *R. aureum*, grows from California to British Columbia, east to the Rocky Mountains and New Mexico. See also Purple Berries.

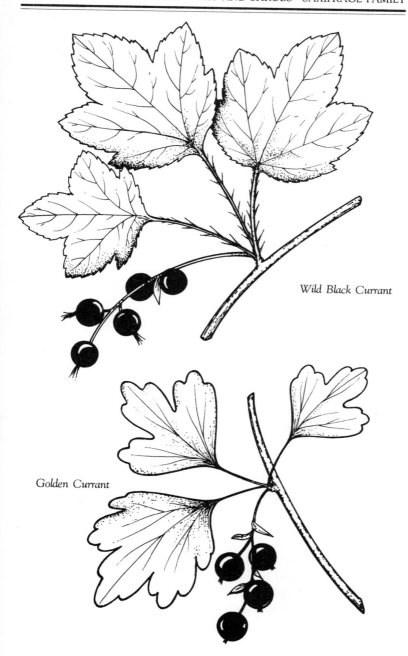

Wild Black Currant

Golden Currant

SMOOTH GOOSEBERRY
Ribes hirtellum

Description: See Purple Berries for general information on gooseberries. This species is an erect shrub, 1 to 3 feet tall, with typical gooseberry leaves. The thorns are short or may be absent altogether. The green or purple flowers are followed by smooth berries, *black or reddish purple*, ripening from June to September.

Habitat: Swamps and moist, rocky woods are favorite locations for Smooth Gooseberry, which grows in northeastern and north central United States.

Berry use: These berries are sweet and tasty. They can be eaten raw or cooked into sauce, jelly, jam or pie.

Similar species: COMMON GOOSEBERRY (*R. inerme*) is a western species that is almost indistinguishable. The berries are *black*. See also Red Berries and Purple Berries.

BLACK CHOKEBERRY
Pyrus melanocarpa

Description: This is a common shrub 2 to 6 feet tall with branchlets and leaves that are smooth. The leaves are elliptic, fine-toothed, pale underneath, with small, dark glands along the midrib on the upper side. White or purplish flowers, 5-petaled, grow in flat-topped clusters. The berrylike pomes are *black*, about ¼ inch in diameter. They ripen in September and October.

Habitat: Black Chokeberry grows in swamps, thickets and clearings or on bluffs. It is common in coniferous forests. It ranges from Newfoundland to Ontario, south to Minnesota, Pennsylvania and the mountains of Georgia.

Use of berrylike fruit: Chokeberries can be used like blueberries. *Warning:* The position of the fruits is very important in Black Chokeberry. Although they are good to eat, they resemble buckthorn berries, which are strongly laxative. The berries of buckthorns are clustered along the twigs. Chokeberry fruit clusters branch off from a separate leafstalk.

Smooth Gooseberry

Black Chokeberry

Another important difference is that Black Chokeberries have a prominent calyx lobe on top and buckthorn berries do not. See Common Buckthorn, p. 192.

JUNEBERRIES
(Rose Family) See Purple Berries.

BLACK RASPBERRY
Rubus occidentalis

Also called BLACKCAP RASPBERRY, this shrub is much like Red Raspberry (see Red Berries), except that it has heavily whitened canes and strong, hooked prickles. The canes arch completely over and may root at the tip. The berrylike fruits are *purplish black*. This species is sometimes called THIMBLEBERRY because when the berry is picked it separates easily from the receptacle, leaving a hollow, so that the berry is thimble-shaped. The range of this shrub is from New Brunswick to North Dakota, south to Georgia, Colorado and New Mexico. The fruits can be used like Red Raspberries.

COMMON DEWBERRY
Rubus flagellaris

Other names: NORTHERN DEWBERRY, PRICKLY DEWBERRY

Description: Closely related to blackberries and raspberries, this trailing, vinelike shrub has scattered, stout, curved prickles. The leaves are thin and dull, light green and toothed. There are 3 to 5 oval or egg-shaped leaflets to a leaf. The white flowers may be an inch across. The fruits *resemble blackberries* in color and shape. They ripen from May to August.

Habitat: Dewberries grow in dry, open places, often in poor soil, from Maine to Minnesota, south to Florida and Texas.

Remarks: There are many species of dewberries. They can be distinguished from blackberries by the fact that blackberry shrubs are more upright.

Use of berrylike fruit: The fruits are sweet and juicy, with a wine flavor, and can be used like raspberries or blackberries. Pectin should be added for jelly or jam.

Northern Dewberry

Highbush Blackberry

Similar species: SOUTHERN DEWBERRY (*R. trivialis*) and SWAMP DEWBERRY (*R. hispidus*) are semi-evergreen shrubs with *black* fruits. The latter has leaves that are dark and shining and stems that are bristly.

BLACKBERRY
Rubus allegheniensis
Other names: HIGHBUSH BLACKBERRY

Description: This shrub has erect or arching, angled, purplish red canes 3 to 6 feet high. It is armed with stout prickles. The leaves have prickly leafstalks and usually 5 egg-shaped leaflets, arranged like a fan. They are toothed, relatively smooth above, densely downy underneath. The white flowers are large, an inch across and in clusters. The "berries" are actually a cluster of drupelets. They are green and hard at first, turning red or purple, then *black at maturity.* They ripen in July or August. Blossoms and fruits may appear on the canes at the same time.

Habitat: Blackberries like lots of sun and coarse, dry soil or rich bottom lands. They grow at the edge of woods, in clearings and thickets. One species or another can be found in most areas of the United States.

Use of berrylike fruit: Blackberries can be eaten raw, cooked into sauce or made into pie, juice, syrup or wine. They are somewhat seedy for jam but make excellent jelly.

Similar species: One authority lists 122 species of blackberries. Blackberry kin include loganberries, a man-made cross of blackberries and raspberries; and boysenberries, a man-made triple cross of blackberries, raspberries and dewberries.

CHERRIES AND PLUMS

See Red Berries for a general description. Cherries and plums with black fruits include the following:

SAND CHERRY (*Prunus pumila*) is a shrub 1 to 6 feet tall with white flowers and *purplish black* fruit, about ⅜ inch in diameter. It is found from southern Canada to New York and Minnesota.

APPALACHIAN CHERRY, also called EASTERN DWARF CHERRY (*P. susquehanae*) is a shrub 2 to 8 feet high with white flowers and *purplish*

Appalachian Cherry

black fruit, about ⅜ inch in diameter. It ranges from Quebec and Maine to North Carolina, west to Illinois and Minnesota.

INDIAN PLUM (*Oemleria cerasiformis*), also called OSO BERRY, is a shrub or small tree 3 to 18 feet tall with ¾-inch oblong fruit, *bluish black or purple*. Although it has a bitter taste, it is edible and makes good jelly. It is found from British Columbia to California.

WOODY VINES AND SHRUBS • KOEBERLINIACEAE

ALLTHORN
Koeberlinia spinosa

Other names: CRUCIFIXION THORN, CORONA DE CRISTO, CROWN-OF-THORNS. There are two other plants bearing the name CRUCIFIXION THORN.

Description: This shrub or small tree is dramatically different, since at times it appears to bear only numerous thorns which grow at right angles to the branches. It is a bushy, green barked shrub growing to 25 feet tall, with leaves reduced to such small scales that it looks leafless. The white flowers bloom in small racemes. The fruit is a *black*, shiny berry.

Habitat: Dense, low thickets are sometimes formed by this shrub, which grows in the deserts of Arizona, New Mexico and Texas.

Remarks: This shrub is of little or no use to livestock. Jack rabbits browse on the twigs.

Berry use: None known.

WOODY VINES AND SHRUBS • CROWBERRY FAMILY

BLACK CROWBERRY
Empetrum nigrum

Other names: HEATHBERRY

Description: This low, creeping, evergreen shrub resembles heather. Its bushy-branched stems spread to form a ground-hugging mat. The short, glossy leaves are dark green and needlelike, crowded along the branches alternately or in whorls. Small, scattered, purplish flowers appear at the

Allthorn

base of the leaves. The round, pea-sized, berrylike fruits turn *purple, then black.* They ripen from July to November and often overwinter.

Habitat: Crowberries grow in peaty soil, rocky coastal areas and gravelly ridges and moraines. They can be found from the Arctic south to northern California and along the northern tier of states to the Atlantic.

Use of berrylike fruit: The fruits are juicy and can be eaten fresh or cooked, but because there are so many seeds, they are usually cooked. Some people do not care for their flavor; this improves after the fruits are frozen. With the addition of pectin, they make excellent jelly. They serve as an important survival food for the Eskimos. Ruffed grouse, ptarmigans and some 40 other species of birds eat the fruits. See also Purple Crowberry, Red Berries.

WOODY VINES AND SHRUBS • HOLLY FAMILY

LOW GALLBERRY HOLLY
Ilex glabra

An evergreen holly, this species has blunt-tipped, leathery leaves that are notched near the tip but are otherwise toothless or possessing only a few wavy teeth above the middle of the leaf. The shrub is sometimes called INKBERRY, a name also given to Pokeweed. The berries are *black* and are eaten by numerous birds, including bobwhites and turkeys, but are *poisonous* to humans, especially children. See Hollies, Red Berries. LARGE GALLBERRY (*I. coriacea*) is similar but larger.

WOODY VINES AND SHRUBS • BUCKTHORN FAMILY

COMMON BUCKTHORN
Rhamnus cathartica

Other names: HARTSHORN, WAYTHORN, RHEINBERRY

Description: This is a medium-sized to large shrub with twigs ending in sharp spines. The leaves are elliptic, pointed and finely toothed and are usually opposite one another. Small, greenish flowers grow in the leaf axils. The drupes are berrylike and *black.*

Black Crowberry

Low Gallberry Holly

Habitat: This shrub was originally cultivated as a hedge and is still used for that purpose, but it has spread to woods and thickets from Nova Scotia to Ontario, south to North Dakota, Missouri and Virginia.

Use of berrylike fruit: Caution is urged. These bitter berries have a strong laxative action. Cases have been reported of children being *poisoned* by eating the fruits.

CAROLINA BUCKTHORN
Rhamnus caroliniana

Other names: INDIAN-CHERRY

Description: This deciduous shrub or tree sometimes grows to 30 feet. It is thornless, as are all buckthorns except Common Buckthorn. The leaves are alternate and simple, lustrous above, paler beneath, prominently veined. Small, yellowish green flowers cluster in the leaf axils. The round, *black,* berrylike fruits are about ⅓ inch in diameter. They ripen in September or October.

Habitat: Carolina Buckthorn grows along streams and on wooded hillsides from Virginia through the Ohio Valley and Nebraska, south to Florida and Texas.

Use of berrylike fruit: Although there are reports of this fruit being cooked into sauce or jelly, *caution* is urged because of its laxative action. It is best regarded as inedible.

Similar species: There are about 100 species of buckthorn. Among those with black fruits are ALDERLEAF BUCKTHORN (*R. alnifolia*), LANCELEAF BUCKTHORN (*R. lanceolata*), EUROPEAN BUCKTHORN (*R. frangula*) and COFFEEBERRY (*R. californica*). See Blue Berries for Supplejack.

PEPPER VINE, WILD GRAPE and FOX GRAPE
(*Vine Family*) See Purple Berries.

WOODY VINES AND SHRUBS • GINSENG FAMILY

HERCULES'-CLUB
Aralia spinosa

Other names: DEVIL'S-WALKINGSTICK

Common Buckthorn

Hercules'-Club

Description: A somewhat grotesque, very spiny shrub or small tree, Hercules'-Club has large leaves that are composed of numerous double-compound leaflets. The trunk and twigs have many stout, coarse prickles. The toothed leaflets are arranged in pairs, except for a terminal leaflet. Small, white flowers grow in terminal clusters. The berrylike fruits are very showy, small, *black* berries on orange stems.

Habitat: Hercules'-Club grows in rich woods and clearings from New England to Iowa, south to Texas and Florida.

Use of berrylike fruit: The fruits are *poisonous if eaten raw.* They are juicy, however, and can safely be cooked into jelly. They are popular with birds.

Similar species: BRISTLY SARSAPARILLA (*A. hispida*) does not grow as far south.

ALTERNATE-LEAF DOGWOOD
(Dogwood family) See Blue Berries.

WOODY VINES AND SHRUBS • HEATH FAMILY

MANZANITA
See Red Berries.

ALPINE BEARBERRIES
See Purple Berries.

BLUEBERRIES, HUCKLEBERRIES, BILBERRIES and DEER-BERRIES
Vaccinium spp. and Gaylussacia spp.

For a general description, see Blue Berries. Other species will be found in the Purple Berries and White or Green Berries sections. Here is a sampling of species with black berries.

SMALL BLACK BLUEBERRY
Vaccinium tenellum

This is one of several blueberry species that has *black*, rather than

Sparkleberry

blue berries. It is also called SLENDER BLUEBERRY. The berries are edible, sweet but rather dry. The range is from Virginia to Florida and west to Mississippi.

SPARKLEBERRY
Vaccinium arboreum

Other names: FARKLEBERRY, TREE HUCKLEBERRY, TREE SPARKLEBERRY

Description: This is a sprawling shrub or small, crooked tree 4 to 30 feet tall. It is deciduous except in the South, where it is an evergreen. Its leaves are oval to oblong, somewhat leathery, glossy above and paler, sometimes downy, underneath. The flowers are white and bell-shaped. They are followed by shiny *black* berries, about ¼ inch in diameter, ripening in September or October.

Habitat: You will find Sparkleberry in dry, sandy soil or rocky woods from Virginia southwest to Oklahoma and south to Texas and Florida.

Berry use: The berries are rather dry and insipid but edible.

BLACK HUCKLEBERRY
Gaylussacia baccata

Other names: TRUE HUCKLEBERRY, WHORTLEBERRY, HIGHBUSH HUCKLEBERRY

Description: This is the most widespread and often the most common huckleberry. It is a beautiful shrub, up to 3 feet tall. It has grayish brown branches and elliptic or egg-shaped leaves. The leaves are covered on both sides with yellow resin-dots. Greenish flowers bloom in short clusters. The berries, which ripen from July to September, are shiny and *black*.

Habitat: Black Huckleberry grows in dry or moist woods, thickets and clearings from Newfoundland to Saskatchewan, south to New England, Georgia and Louisiana.

Berry use: The berries are abundant but neglected, perhaps because people fear they may be poisonous. They have a delicious flavor, somewhat tangy and sweet. Because of their seeds, they are best cooked and strained to remove the seeds, then eaten as sauce or made into jelly.

Black Huckleberry

WOODY VINES AND SHRUBS • SAPODILLA FAMILY

EASTERN BUMELIA
Bumelia lycioides

Other names: BUCKTHORN BUMELIA, a confusing name, since it does not belong to the buckthorn family.

Description: This deciduous shrub or small tree sometimes grows 20 feet or more in height. Its leaves are alternate, narrowly elliptic, untoothed, smooth, or nearly so, on both sides. The branchlets have short thorns and milky sap. Small white flowers are borne in dense clusters in the leaf axils. The fruits are *black* and berrylike. They ripen from September to November.

Habitat: Eastern Bumelia can be found in swampy woods, and on bluffs, stream banks and dunes from Virginia to Florida, west to Texas and north to Illinois.

Use of berrylike fruit: The bittersweet pulp can be eaten raw.

WOOLLY BUMELIA
Bumelia lanuginosa

Other names: GUM BUMELIA, WOOLLY BUCKTHORN, GUM-ELAS-TIC, CHITTAMWOOD

Description: This species is similar to the preceding, but its leaves are rusty-hairy beneath and are more or less evergreen. Its fruits are *black*, round or oval, about ½ inch long. They ripen about October.

Habitat: Woolly Bumelia grows in the Southeast and north in the Mississippi Valley to Illinois. It also grows in Texas and New Mexico.

Remarks: The sap of this shrub is sometimes used as chewing gum.

Use of berrylike fruit: The raw fruit is edible.

WOODY VINES AND SHRUBS • HONEYSUCKLE FAMILY

BLACK TWINBERRY
Lonicera involucrata

Other names: INVOLUCRED FLY HONEYSUCKLE, BEARBERRY HON-EYSUCKLE, TWINBERRY

Woolly Bumelia

Description: See Red Berries for a description of honeysuckles. This species has squarish branches, yellow flowers and *black* berries, round or oval, paired on long axillary stalks.

Habitat: Black Twinberry grows in cool woods or along streams from New Brunswick to British Columbia and Alaska, south to Mexico and Michigan. It is one of the most common honeysuckles in the West.

Berry use: The berries can be eaten raw but are not appetizing. They are better cooked, dried or made into jelly or syrup.

JAPANESE HONEYSUCKLE
Lonicera japonica

This vine forms dense tangles as it climbs over underbrush or sprawls on the ground. The upper leaves are not united; the lower ones may be lobed. The flowers are fragrant and lovely to behold, white or yellow, tinged with red or purple. The *purplish black* berries ripen from September to November. This plant, introduced from Asia, has become a problem because of the speed with which it spreads, covering large areas of plants and saplings and damaging them by cutting off their sunlight. See Red Berries and Blue Berries for other honeysuckles.

WOLFBERRY
See White or Green Berries.

VIBURNUMS
Viburnum spp.

Viburnums are common and widespread in eastern United States, but the name is not well known because most of the more than 20 species have popular names that do not include the word "viburnum." The flowers are white or pinkish and have a short-tubed corolla with 5 lobes. The fruits are small, fleshy, berrylike drupes, usually *black,* containing a single flat seed. Deer, beavers, rabbits, chipmunks, squirrels, mice, skunks, game birds and songbirds eat the fruits. Here are some of the more common viburnums with black fruits:

Black Twinberry

HOBBLEBUSH
Viburnum alnifolium

Other names: WITCH-HOBBLE, MOOSEWOOD

Description: This straggly shrub, 3 to 10 feet high, has forked branches that often bend down and root at the tip. They trap the unsuspecting intruder with these loops, or "hobbles." The twigs, buds and undersides of leaves are covered with rusty hairs. The large leaves are roundish or heart-shaped, broadly pointed and finely toothed. Hobblebush flowers are of two kinds, forming a flat cluster 3 to 5 inches across. The outer flowers, which are sterile, are large and showy, the inner ones are minute. The fruit turns *red, then blue, then black* when it is mature.

Habitat: This species is found in open areas or woods from southern Canada to northern United States, and in the mountains to Georgia.

Use of berrylike fruit: When ripe the fruit can be eaten raw. It can also be cooked into sauce or, with pectin added, jelly.

Similar species: WAYFARING TREE (*V. lantana*) is similar but has smaller leaves.

BLACKHAW
Viburnum prunifolium

Other names: STAGBUSH, CRAMP BARK, STAGBERRY, SMOOTH BLACKHAW

Description: This shrub or small tree, 5 to 15 feet tall, has stiff, horizontal branches and checkered bark. The small, oval leaves are broadly pointed and have fine teeth. Large, flat clusters of white flowers, 3 to 4 inches across, bloom before the leaves unfurl. By September or October, the *blue-black*, oval berries are ripe.

Habitat: Blackhaw thrives along roadsides and in abandoned fields and pastures from New York and Connecticut west to Kansas and south to Florida and Texas.

Remarks: The name "Blackhaw" is derived from the fact that its berries are black and its stout spurs resemble those of the hawthorn.

Use of berrylike fruit: The fruit has a sweet, pleasant taste and can be eaten raw. It becomes sweeter after the first frost. Some people don't

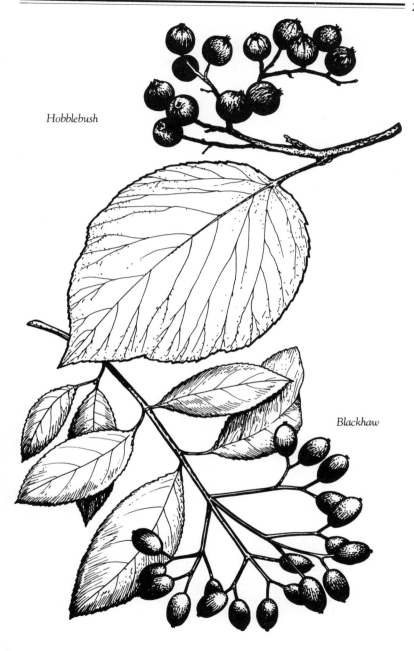

Hobblebush

Blackhaw

care for the flavor of the cooked fruit; others use it for jelly, jam or sauce.

DOWNY VIBURNUM
Viburnum rafinesquianum

Other names: DOWNY ARROWWOOD, SHORTSTALK ARROWWOOD

Description: This species is small, 2 to 5 feet high, with egg-shaped leaves that are smooth above, downy underneath. The flat-topped cluster of flowers is followed by oval, berrylike fruits. They are *purplish black or deep purple,* with a whitish bloom.

Habitat: The range of this species is from Quebec to Manitoba, south to North Carolina, Georgia, Kansas and Missouri.

Use of berrylike fruit: The fruit of this species is bitter and seldom used.

WILD-RAISINS
NORTHERN, *Viburnum cassinoides;* SOUTHERN, *Viburnum nudum*

Other names: NORTHERN and SOUTHERN WITHEROD, POSSUM HAW (*V. nudum*), APPALACHIAN TEA (*V. cassinoides*)

Description: The northern species grows from 3 to 8 feet tall, the southern 5 to 15 feet. The leaves are narrowly egg-shaped. Small white flowers, grouped in flat-topped clusters, are followed by fruits that are roundish or slightly oval. They are *at first yellow or red, then blue, finally blue-black,* with a white bloom. They often hang on to the bush far into the winter.

Habitat: Wild-Raisins thrive in wet woods, swamps and bogs. The range of the northern variety is from Newfoundland to Ontario, south to the mountains of Alabama and the Great Lakes area. The southern variety grows from Connecticut to Kentucky and Arkansas, south to Florida and Texas.

Use of berrylike fruit: The thin pulp of the ripened fruits has a flavor like a cross between raisins and dates, and is good raw. With the seeds removed, the fruits can be used for sauce or jelly, with pectin added. They are best mixed with a fruit that has a more tart flavor. When dried the fruits look like raisins.

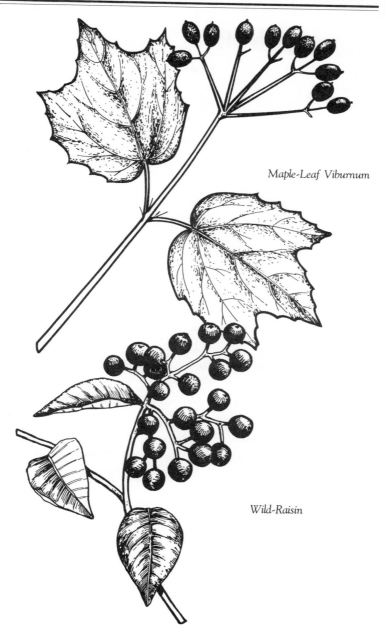

Maple-Leaf Viburnum

Wild-Raisin

MAPLE-LEAF VIBURNUM
Viburnum acerifolium

Description: The broad, 3-lobed leaves of this small shrub resemble those of maple trees. In the fall they turn pinkish to magenta and are very attractive. The fruits are round or slightly oval-shaped, *glossy black* with a bloom when mature.

Habitat: This species ranges from Quebec to Minnesota, south to Georgia and Mississippi.

Use of berrylike fruit: There is no mention of this species being used by humans.

NANNYBERRY
Viburnum lentago

Other names: SHEEPBERRY, SWEET VIBURNUM

Description: This species occasionally attains a height of 30 feet but is more often 5 to 15 feet. Its twigs are long and flexible, and the stalks of the upper leaves are "winged." The berrylike fruits are *bluish black* at maturity, whitened with a bloom. They are borne on slender, reddish stems.

Habitat: Nannyberry grows in bottom lands and rich woods from Quebec to Manitoba, south to Georgia, Mississippi and Colorado.

Remarks: The wood of this species has a disagreeable odor.

Use of berrylike fruit: The fruits are sweet and can be eaten raw. They are also good cooked in sauce or preserves, but you may wish to remove the seeds. The fruit hangs on the shrub in winter and is not harmed by the frost.

ARROWWOOD
Viburnum dentatum

Description: This is a bushy shrub 3 to 10 feet tall. The opposite leaves, egg-shaped or roundish, have prominent veins ending in large, sharp-pointed, marginal teeth. The flowers grow in typical viburnum clusters. From July to October, the berrylike fruits, *bluish black,* ¼ to ⅜ inch across, ripen.

Nannyberry

Habitat: This viburnum grows in swamps and woods from Nova Scotia to Ontario, south to Florida and Texas.

Remarks: Arrowwood is a variable species, divided into two or more species by some botanists. The name comes from the fact that Indians trimmed straight young shoots of this shrub into arrow shafts.

Use of berrylike fruit: If the seeds are removed, the fruits make a fine sauce. If pectin is added, jelly can be made with the juice.

COMMON ELDERBERRY
Sambucus canadensis

Other names: ELDER, AMERICAN or SWEET ELDER

Description: Common Elderberry is a deciduous shrub that grows 4 to 12 feet high or more and sometimes spreads into large patches by underground runners. The compound leaves are large, toothed, opposite, with a leaflet at the tip. They are paler and slightly downy underneath. Tiny, star-shaped, white flowers bloom in large, flat-topped clusters measuring to 8 inches across. They are showy and lacy in appearance. The berries ripen from August to October, often bending the branches with their weight. They are *usually purplish black,* but they go through *stages of red, blue and purple.* They are small and juicy.

Habitat: This shrub likes rich damp soil and is often found along stream banks, in roadside ditches, thickets and open woods. This species grows throughout the eastern half of the United States.

Remarks: Flutelike whistles are easy to make from the pithy stems of elderberry. Indians, who made such flutes, called the elderberry the "tree of music." They also chose long, straight stems, aged them and made arrows from them.

Berry use: Unripe berries are poisonous and can cause digestive upsets. Harvest only when the berries are blue, purple or black. The similar Red-Berried Elder (see Red Berries) is reportedly poisonous even when the berries are ripe and cooked. Neither the smell nor the taste of raw elderberries is very appetizing, but ripe berries have many uses when properly prepared. They are easy to harvest by running fingers over the stems. The juice can be used to make excellent drinks (alcoholic or nonalcoholic) or jelly, with pectin added. The berries may be dried,

Arrowwood

cooked into a sauce or used in pie fillings or muffins. Cardinals, bluebirds, tanagers, thrushes, vireos, wrens, robins, cedar waxwings, grouse, pheasants and pigeons enjoy the berries. Bears are also fond of them.

Similar species: Western species of elderberry include BLUEBERRY ELDER or BLUE ELDER *(S. glauca* or *caerulea),* with *blue* berries; VELVET ELDER *(S. velutina),* with *blue-black* berries; PACIFIC RED ELDER *(S. callicarpa),* also called CALIFORNIA ELDERBERRY, with *red* berries; BLACK BEAD ELDER *(S. melanocarpa),* also called BLACK ELDERBERRY, with *black* berries.

TREES • PALM FAMILY

CABBAGE PALMETTO
Sabal palmetto

Other names: SWAMP CABBAGE

Description: This evergreen tree has fan-shaped leaves clustered at the top. The stems are stiff and half-rounded, 6 to 7 feet long. The leaf blades are lustrous and dark green, divided into numerous ribbonlike segments that are long and drooping and have many threadlike filaments. The fragrant flowers are small and white, in drooping clusters 2 to 2½ feet long. The fruits are drupelike berries (1-seeded), round to oval, *nearly black* when they ripen in the fall.

Habitat: Cabbage Palmetto is found from North Carolina throughout Florida on coastal plains.

Berry use: When ripe, the berries are good eaten raw or made into syrup.

Similar species: CALIFORNIA PALM *(Washingtonia filifera),* also called DESERT PALM, is much like Cabbage Palmetto except that the dead leaves commonly hang on the tree, forming a thatchlike, drooping mass. The drupelike berries are *black,* resembling small grapes and hanging in loose clusters high in the tree. This palm tree is found in southern California and southwestern Arizona. The berries may be eaten raw, roasted, or made into a syrup.

REDBAY
(Laurel Family) See Blue Berries.

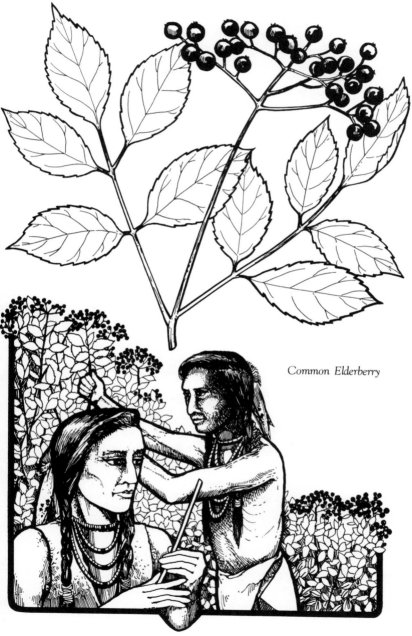

Common Elderberry

TREES • ROSE FAMILY

HAWTHORN
See Red Berries.

BLACK CHERRY
Prunus serotina

Other names: AMERICAN BLACK CHERRY, WILD BLACK CHERRY, RUM CHERRY, SWEET BLACK CHERRY, CABINET CHERRY

Description: This medium-sized to large tree grows to 100 feet or even more. The leaves are shiny, long, narrow and blunt-toothed, with the midrib prominently fringed with hairs underneath. White flowers appear with the leaves and are borne in long, drooping spikes. The fruits ripen in late summer and early fall. They are round, about ½ inch in diameter. Their color changes with maturity from *bright red through deep purple to glossy purplish black.*

Habitat: Black Cherry trees are usually found in mixed stands in rich, moist areas throughout the East to Minnesota and Texas.

Remarks: The name "Rum Cherry" resulted from the use of the sweetened juice of this tree as an addition to rum or brandy to produce a mellower flavor. The tree's heartwood is reddish brown and greatly prized for furniture.

Use of berrylike fruit: The pulp of this cherry can be enjoyed raw, though it is somewhat tart. It is excellent for jelly, with pectin added, and for sauce and pie. Songbirds and game birds enjoy the fruit, and animals keep what falls on the ground from going to waste.

Similar species: CAROLINA LAUREL CHERRY (*P. caroliniana*), also called LAUREL CHERRY, MOCK-ORANGE, and WILD-ORANGE, is an attractive evergreen tree with an oblong, open crown. The leaves have a pleasant, cherrylike odor when crushed. Often planted as an ornamental, this tree grows from North Carolina to Florida and Texas. Birds like the *black* (or nearly black) fruits, but humans are not fond of them, except when they are used for jelly or jam.

WESTERN SOAPBERRY
(*Soapberry Family*) See Yellow or Orange Berries.

Cabbage Palmetto

BLACK TUPELO

(Tupelo Family) See Blue Berries.

TREES • VERBENA FAMILY

FLORIDA FIDDLEWOOD

Citharexylum fruticosum

Description: This is an evergreen tree or shrub that grows to 30 feet tall. Its oblong leaves are simple, opposite, 3 to 4 inches long, with curled margins. The small, white flowers, in axillary clusters, are fragrant and bloom all year. The fruit is a glossy, *reddish brown to black* drupe, ⅓ of an inch in diameter.

Habitat: This tree grows only in the southern half of Florida.

Use of berrylike fruit: The fruit is edible and juicy, but some people do not care for the flavor.

Black Cherry

Florida Fiddlewood

Some berries are on the borderline between reddish orange and orangish red. Consult Red Berries if you don't find what you're looking for here.

FLOWERING HERBS • ARUM FAMILY

GREEN DRAGON

See Red Berries.

FLOWERING HERBS • LILY FAMILY

FAIRYBELLS

See Red Berries.

FLOWERING HERBS • BARBERRY FAMILY

MAY APPLE
Podophyllum peltatum

Other names: MANDRAKE, AMERICAN MANDRAKE, WILD MANDRAKE, HOG APPLE, WILD LEMON, RACCOON BERRY

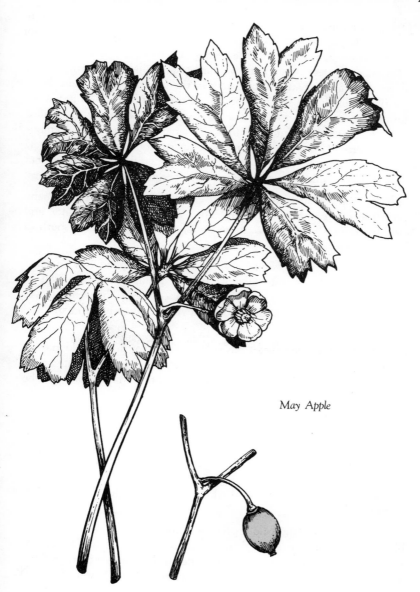

May Apple

Description: Erect stems, 1 to 1½ feet high, spring from a perennial, horizontal, poisonous rootstock. There are two kinds of stems, a nonflowering stem ending in the center of a basal leaf and a flowering stem that divides to support two large, umbrellalike, 5- to 9-lobed leaves. The flower hangs from a stalk in the fork where the stems bearing the umbrellalike leaves join. It is a single, nodding, white, waxy flower, 1½ to 2 inches broad. Its odor strikes some people as fragrant and others as unpleasant. The fruit is a large, fragrant, *yellow*, lemon-shaped berry, 12 to 18 inches long. It ripens from July to September, when the rest of the plant is dying. It is so heavy that it is not unusual to find it has pulled the plant almost to the ground.

Habitat: These plants often grow in colonies so dense that for many yards the earth is nearly hidden. They are found in woodlands and clearings from Minnesota, southern Ontario and New England south to Texas and Florida.

Remarks: In spite of its being called Mandrake, this plant is not related to the Mandrake of the ancients, which Roman surgeons used to deaden pain 1,800 years before ether was discovered. The stems, leaves and seeds are poisonous and the root is a violent purgative. Indians used the plant as a laxative.

Berry use: The green berries are strongly laxative and should not be eaten. The pulp around the seeds of the ripe berries can be eaten raw, except by children. With the addition of pectin, May Apples can be used to make jelly.

FLOWERING HERBS • ROSE FAMILY

CLOUDBERRY
Rubus chamaemorus

Other names: BAKE-APPLE, BAKED APPLE BERRY, MOUNTAIN RASPBERRY

Description: This plant belongs to the same family as raspberries and blackberries, but it is an herb rather than a shrub. It has a creeping stem with erect shoots from 3 to 8 inches high. There are no prickles. The leaves are broad and 5-lobed. The white, solitary flowers rise from an unbranched stem. Ripening in July or August, the berrylike fruits are *golden, amber or pale red.*

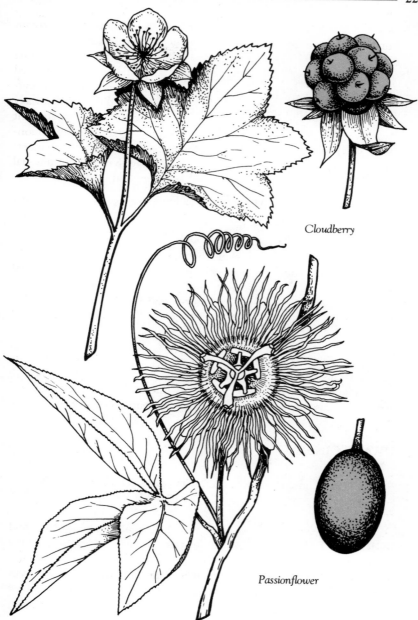

Cloudberry

Passionflower

222

Habitat: Cloudberries grow in peat bogs, swamps and tundra from Maine and New Hampshire to the Arctic and west to British Columbia and Alaska.

Use of berrylike fruit: Cloudberries are delicious fresh, with cream and sugar, or cooked into sauce, jam, jelly or a pie. They are also used to make a liquor with a distinctive flavor.

FLOWERING HERBS • PASSIONFLOWER FAMILY

PASSIONFLOWER
Passiflora incarnata

Other names: MAYPOP, APRICOT-VINE, PASSION VINE

Description: This weak, trailing or climbing vine with tendrils has a spectacular flower. Up to 3 inches wide, it is fragrant, solitary, with alternating white petals and sepals that are crowned with a fringed halo of long, pink or purple, delicate filaments. The large, alternate leaves have 3 to 5 lobes. The fruit is an oval *yellow* berry about the size of a hen's egg. It ripens from July to October and is called a maypop.

Habitat: Look for this flowering plant in thickets or fields from Pennsylvania to Oklahoma, south to Texas and Florida.

Berry use: The berries are edible and can be eaten raw or made into jelly, with pectin added, or into a drink by simmering with lemon and sugar. Their juice is said to be good for whitening teeth. Formerly, the berry juice was used for sore eyes. Deer and birds eat the berries.

DWARF GINSENG
Ginseng Family

This is like Ginseng, only smaller, with *yellow* berries. See Red Berries.

FLOWERING HERBS • NIGHTSHADE FAMILY

HORSE-NETTLE
Solanum carolinense

Other names: BULL NETTLE, DEVIL'S-TOMATO, DEVIL'S-POTATO, SAND-BRIER

Horse-Nettle

Description: This is a prickly plant growing 1 to 4 feet tall. It has widely toothed leaves. The flowers are violet or white, star-shaped, with the stamens forming a cone in the center. The berries are *yellow or orange*, tomatolike, the size of a cherry.

Habitat: Horse-Nettle grows in gardens, fields and waste ground from southern New England west to Minnesota and Nebraska, south to Florida and Texas.

Remarks: The plant's leaves are eaten by the tomato beetle, and because this helps keep these beetles alive, it is regarded as an undesirable weed.

Berry use: The berries are *poisonous.* See Bitter Nightshade in Red Berries.

Similar species: WHITE HORSE-NETTLE (*S. elaeagnifolium*), also called SILVER HORSE-NETTLE, SILVERLEAF HORSE-NETTLE or BULLNETTLE, has lavender, starlike flowers and *yellow* berries. It grows in dry, open areas in the Southwest to central California.

CLAMMY GROUND-CHERRY
Physalis heterophylla

This plant is similar to Virginia Ground-Cherry (see Red Berries). The stems are sticky as well as hairy; the leaves have rounded bases and few teeth. The berry is *yellow* and edible. It is found in dry woods and clearings from Saskatchewan to New England and south. STRAWBERRY TOMATOES (*P. pruinosa* and *fendleri*) are ground-cherries that also have *yellow* berries, edible when ripe. They are found from the East Coast to Arizona.

BITTER NIGHTSHADE
See Red Berries.

FLOWERING HERBS • HONEYSUCKLE FAMILY

FEVERWORT
Triosteum perfoliatum

Other names: TINKER'S-WEED, HORSE GENTIAN

Description: This plant, 2 to 4 feet tall, is notable for the way the wide

Feverwort

Wild Coffee

stalks of the paired leaves meet and surround the stem. Flowers are located in the leaf axils. Green, yellow or dull purplish brown, they are bell-shaped and embraced by 5 long sepals. The berries, maturing from August to October, are hairy, *orange to reddish orange*, clustered in the node at the stem. They are crowned by 5 long sepals and contain 3 nutlets.

Habitat: Feverwort grows in dry to medium woods in Minnesota, Wisconsin, Michigan and Massachusetts and southward.

Berry use: The ripe berries can be dried, roasted, ground, and then used to make a hot drink.

Similar species: WILD COFFEE (*T. aurantiacum*) has leaves that taper to their bases. Otherwise, it is similar in appearance, habitat and berry use.

WOODY VINES AND SHRUBS • ROSE FAMILY

FIRETHORN
Cotoneaster pyracantha

Description: This shrub, growing to 10 feet tall, is usually an evergreen. It has thick leaves with wedge-shaped bases, rounded tips and wavy edges. Sharp purple spines are present. The small white flowers grow in clusters. The round, 5-seeded berries, are fiery *orange or red* and in clusters. They usually remain on the shrub all winter.

Habitat: This is a European shrub that has escaped cultivation from Pennsylvania to Florida and Louisiana.

Berry use: It is not known whether the fruits are edible.

THIMBLEBERRY
See Red Berries.

WILD ROSE
See Red Berries.

AMERICAN, CANADA, CHICKSAW, HORTULAN and WILDGOOSE PLUM
See Red Berries.

Firethorn

CHRISTMAS BERRY
See Red Berries.

PINCUSHION CACTUS
(*Cactus family*) See Red Berries.

HOLLY
(*Holly Family*) See Red Berries.

AMERICAN BITTERSWEET
(*Bittersweet Family*) See Red Berries.

WILD GRAPE
(*Vine Family*) See Purple Berries.

WOODY VINES AND SHRUBS • OLEASTER FAMILY

CANADA BUFFALOBERRY
Shepherdia canadensis

Description: This is the best known of the three varieties of Buffaloberry native to the United States and Canada. It is deciduous and grows 3 to 10 feet tall. It is the only plant with opposite leaves having undersides densely coated with rusty scales. These scales are also present on the twigs, which are not angled. The leaves are slender, short-stalked, oval and from ¾ to 1½ inches long. Small, yellowish green flowers appear before the leaves are out. They are bell-shaped and grow in a cluster. The berrylike fruits are *yellow, orange or reddish,* small, translucent and egg-shaped. Each contains 1 seed. Ripening time is from June to August.

Habitat: This species grows from Newfoundland to Alaska, south to New England and west to the Rocky Mountains and New Mexico.

Remarks: The name buffaloberry probably results from the fact that Indians frequently served the berries with buffalo meat. This plant is one of the few not belonging to the pea family that has nitrogen-fixing bacterial root nodules present.

Use of berrylike fruit: The fruit is bitter and disagreeable-tasting but edi-

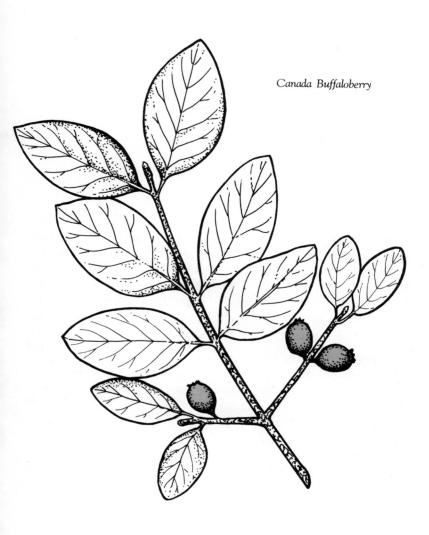

Canada Buffaloberry

ble, and can thus serve as an emergency food. Raw fruits contain saponin, which is used commercially to produce foam. In Alaska certain Indians mix the fruits with sugar and water to make a foam topping which takes the place of whipped cream. Bears, chipmunks, ground squirrels, catbirds, brown thrashers and quail eat buffaloberries. See also Silver Buffaloberry in Red Berries.

DEERBERRY

(*Heath Family*) See White or Green Berries.

WOODY VINES AND SHRUBS • HONEYSUCKLE FAMILY

WILD RAISIN

See Black Berries.

SQUASHBERRY

Viburnum edule

Other names: MOOSEBERRY

Description: This is a sprawling shrub about 2 to 5 feet high. The leaves usually have 3 short, often uneven lobes, and are irregularly toothed. A cluster of white or pinkish flowers, less than 1½ inches broad, blooms from May to August. The berrylike fruits are round or slightly egg-shaped, ranging from *yellow to red*. They ripen from August to October.

Habitat: Squashberries grow in cool, moist woods and ravines from Labrador to Alaska, south to New York and Oregon.

Use of berrylike fruit: The fruits can be eaten raw or used to make excellent jelly, jam or sauce. They are rich in vitamin C and have been used in the treatment of scurvy and other ailments. Squirrels and grouse are fond of them.

ORANGE HONEYSUCKLE

See Red Berries.

COMMON ELDERBERRY

See Black Berries.

Squashberry

TREES • ELM FAMILY

SUGARBERRY
(Elm Family) See Red Berries.

COMMON HACKBERRY
See Purple Berries.

TREES • MULBERRY FAMILY

FLORIDA STRANGLER FIG
Ficus aurea

Description: This is a most unusual tree. Its seeds send forth aerial roots, which reach the ground and establish an independent root system. Sometimes the aerial roots encircle the host tree and "strangle" it. Florida Strangler Fig is an evergreen, 40 to 50 feet tall, with leathery, toothless leaves that are elliptic to oblong. The small flowers are yellow at first, then turn red. The fruit consists of small drupelets imbedded in a fleshy receptacle that looks like a berry. It is *yellow to reddish purple,* about ½ inch in diameter.

Habitat: This tree grows in approximately the southern half of Florida.

Use of berrylike fruit: The fruits are most commonly eaten raw or made into preserves.

TREES • CUSTARD APPLE FAMILY

PAWPAW
Asimina triloba

Other names: COMMON PAWPAW, CUSTARD APPLE, WILD BANANA, FALSE BANANA

Description: A small to medium-sized tree, Pawpaw grows 20 to 50 feet tall. It has a short trunk and a spreading crown, with somewhat twisted branches and dark-brown bark that has whitish blotches and wartlike protuberances. The large leaves stay on the tree 2 to 6 years. They are elliptical, alternate, untoothed, and often drooping. The flowers, which come out before the leaves, are about 2 inches in diameter, with

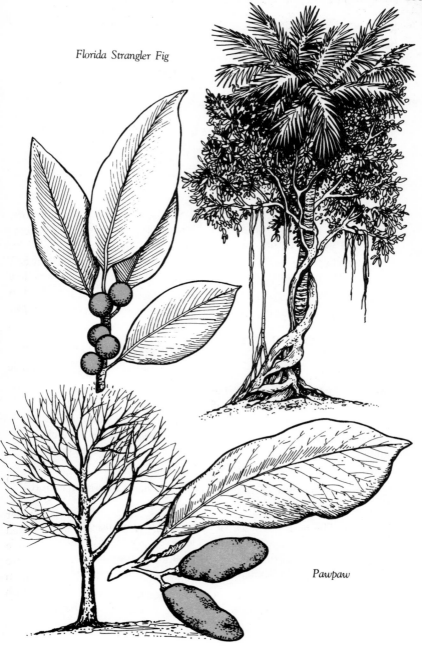

Florida Strangler Fig

Pawpaw

heavily veined purple petals arranged in two sets of 3. The fruit is a berry, bananalike, irregularly oblong, up to 5 inches in length. The color is *greenish yellow to dark brown.*

Habitat: Pawpaw is scattered throughout the eastern United States, growing in rich soil on stream banks and in woods.

Berry use: The pulp of this berry is sweet, yellow and creamy-textured. Some people have to acquire a taste for it. It is best to let it ripen until it has brownish splotches or becomes dark brown before using. It can be used raw or in pies, but caution is advised. Some people develop a skin rash from handling or eating the fruit. Raccoons and opossums are especially fond of Pawpaws.

Similar species: In addition to several species of Pawpaw, the custard apple family includes POND-APPLE *(Annona glabra).* This tree is taller than the Pawpaw, with smaller leaves that are shed late in the fall. The flowers are yellowish white, about an inch in diameter. The fruit is *greenish yellow or brown* and can be eaten raw or made into jelly.

TREES • SOAPBERRY FAMILY

WESTERN SOAPBERRY
Sapindus drummondii

Description: This is a deciduous shrub or small tree growing to 50 feet tall, with scaly, reddish bark. The alternate leaves are compound, with 4 to 9 pairs of lance-shaped leaflets, untoothed, smooth above, hairy below. The white-petaled flowers are very small, blooming in loose, branching clusters. The round berries are *yellow,* about ½ inch in diameter, *turning black* as they dry. They remain on the branches until spring.

Habitat: Western Soapberry grows from Louisiana to Arizona, north to Missouri, Kansas and Colorado.

Berry use: The flesh of the berries lathers in water and can be used as a soap substitute, hence the name. However, some individuals develop skin irritations from the berries. They are not regarded as edible. The hard, black seeds are sometimes drilled and strung as beads.

Similar species: FLORIDA SOAPBERRY *(S. marginatus)* is a small tree that grows from South Carolina to Florida. It also has *yellow* berries.

Pond-Apple

Western Soapberry

BUTTERBOUGH
See Purple Berries.

OREGON CRAB APPLE
Malus diversifolia

Description: A small tree, up to 30 feet tall, with an irregular crown, Oregon Crab Apple grows in dense, spiny thickets. The leaves are dark green above, lighter below. White and fragrant blossoms have 5 petals. The pome is oblong, berrylike in appearance, ½ to ¾ of an inch long, *greenish yellow or reddish purple.*

Habitat: This tree is found in western Washington, Oregon and northern California.

Use of berrylike fruit: The apples are too sour to enjoy raw, but they may be used for jellies and preserves.

AMERICAN MOUNTAIN-ASH and HAWTHORN
(*Rose Family*) See Red Berries.

PACIFIC MADRONE
Arbutus menziesii

Description: This is an interesting tree of the Pacific Coast. It can be recognized by the way its reddish brown bark flakes off in big, thin scales, revealing the light-red inner bark. The branches are thick and curving. It is a medium-sized tree, 20 to 100 feet high and 1 to 4 feet in diameter. The evergreen leaves are alternate, leathery and shiny, dark green above and whitish below, oval to oblong in shape, without teeth. The small, white urn-shaped flowers grow in drooping clusters 5 or 6 inches long. The fruits are round, with a rough surface, berrylike, ½ inch in diameter, in clusters. They may be *yellow, orange or red.* Since they take a year to ripen, they are seen side-by-side with the flowers.

Habitat: This tree is native along the Pacific Coast from British Columbia to southern California.

Use of berrylike fruits: The yellow, mealy flesh of the fruit is edible, but some people find it unpalatable. However, many birds are fond of it.

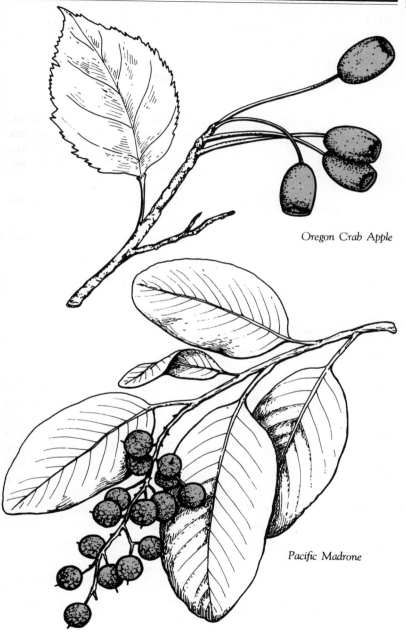

Oregon Crab Apple

Pacific Madrone

Similar species: TEXAS MADRONE *(A. texana)* has fruits *dark red* in color. ARIZONA MADRONE *(A. arizonica)* fruits are *orangish-red.*

TREES • SAPOTE FAMILY

FALSE-MASTIC
Sideroxylon foetidissimum

Other names: JUNGLE PLUM

Description: This medium-sized tree has an irregular crown. Its leaves are alternate, oval and evergreen, 3 to 5 inches long. Inconspicuous yellow flowers bloom in the leaf axils. The berrylike fruit is *yellow,* an inch long, and has thick, juicy flesh.

Habitat: False-Mastic is found along Florida's east coast and in the southern quarter of the state.

Use of berrylike fruit: The fruit is edible raw, but it is acid and bitter.

TREES • EBONY FAMILY

COMMON PERSIMMON
Diospyros virginiana

Other names: BLACK SAPOTE, DATE PLUM

Description: This is a small to medium-sized, thin, deciduous tree 25 to 50 feet high, with black or dark-gray, furrowed bark and a broad, rounded crown. It has alternate, untoothed, oval leaves pointed at the tip. They are dark green and glossy above, pale and sometimes hairy beneath. The small, yellow-green, bell-shaped flowers appear with the leaves. Ripening from September to October, the round berries are *orange to reddish purple,* 1 to 1½ inches long, and wrinkled when fully ripe. The withered, 4-lobed calyx remains attached to the berry. The fruit stays on the tree all winter unless it is harvested.

Habitat: The Common Persimmon tree is native to eastern and southern United States from New York to Missouri, south to Florida and Texas.

Berry use: The berries are too astringent to enjoy when they are green, but after a frost they are sweet, soft and tasty. They can be eaten raw or made into jam, pudding or pie.

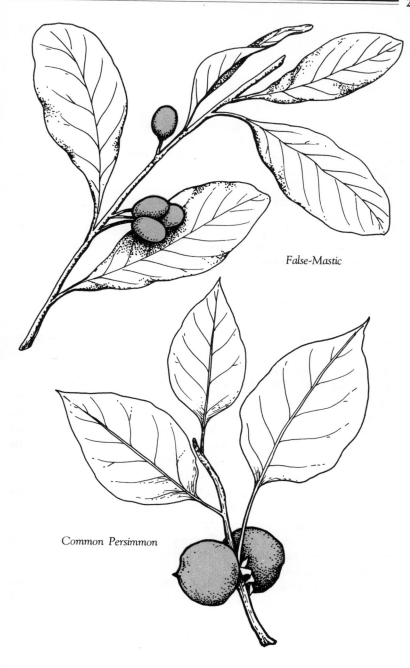

False-Mastic

Common Persimmon

GLOSSARY

achene: A small, dry, hard, one-seeded fruit.

alternate: Not opposite on the stem.

anther: The pollen-bearing part of the stamen.

axil: The angle between the leafstalk and the stem.

axillary: Located in an axil.

berry: A simple fruit which is fleshy and pulpy throughout, with seeds imbedded in the pulp.

bract: A modified leaf situated beneath a flower or a flower cluster.

calyx: The outer whorl of sepals, which may be separate or fused.

compound leaf: A leaf divided into leaflets.

corm: The enlarged, bulblike base of a stem.

corolla: The petals of a flower taken collectively.

double-toothed: Having large teeth which have smaller teeth on them.

drupe: A fleshy fruit with the seed enclosed in a hard, bony covering.

drupelet: A tiny drupe.

egg-shaped leaf: One that is broadest below the middle.

elliptic leaf: One that is widest in the middle and tapers at both ends.

fruit: The seed-bearing part of a plant.

herb: A plant with no persistent woody stem aboveground.

herbaceous: A fleshy, non-woody plant.

involucre: A circle of bracts beneath a flower cluster or fruit.

lance-shaped leaf: A leaf that is much longer than it is wide, tapering to a point, the widest part being near the base.

node: The place on a stem where the leaf is attached.

ovary: The swollen base of the pistil where seeds develop.

palmately compound leaf: A leaf arranged so that the leaflets radiate from one central part.

pinnately compound leaf: A leaf with leaflets arranged along the sides of a central part or midrib.

pistil: The female organ of a flower which develops into the fruit and seeds.

pith: The soft tissue in the center of a twig or stem.

pome: A fleshy fruit, such as an apple, that has a core.

raceme: A long cluster of flowers arranged singly along a stalk, each flower having its separate stalk.

receptacle: The expanded end of a stem which bears the organs of the flower or the fruit.

rhizome: A rootlike, subterranean stem.

semiherbaceous: Only partly woody.

sepal: One of the parts that compose the calyx, surrounding all other parts of the flower.

sessile: Without a stalk.

simple: Not divided into parts.

spadix: A club-shaped stalk on which small flowers are crowded.

spathe: The hooded, leaflike sheath enfolding the spadix.

spike: A long flower cluster with stalkless or near-stalkless flowers along the stem.

stamen: The male part of a flower which produces the pollen.

stipule: A small, leaflike appendage at the base of a leafstalk. It leaves a scar on the twig when it drops.

umbel: An umbrellalike flower cluster in which the flower stalks radiate from the same point.

whorl: Three or more leaves radiating from a single point.

wing: A thin membrane extending from a seed, fruit or leafstalk.

SELECTED BIBLIOGRAPHY

Angier, Bradford, *Feasting Free on Wild Edibles*. Harrisburg, Pa.: Stackpole Books, 1972.

——, *Field Guide to Edible Wild Plants*. Harrisburg, Pa.: Stackpole Books, 1974.

——, *Field Guide to Medicinal Wild Plants*. Harrisburg, Pa.: Stackpole Books, 1978.

——, *Free for the Eating*. Harrisburg, Pa.: Stackpole Books, 1966.

Baerg, Harry J., *The Western Trees*, 2nd ed. Dubuque, Iowa: William C. Brown Co., 1973.

Beedell, Suzanne, *Pick, Cook and Brew*. London: Pelham Books, 1973.

Benson, Lyman, *The Native Cacti of California*. Stanford, Calif.: Stanford University Press, 1969.

———, and Robert A. Darrow, *The Trees and Shrubs of the Southwestern Deserts*. Tucson: University of Arizona Press, 1954.

Berglund, Berndt, and Clare E. Bolsby, *The Complete Outdoorsman's Guide to Edible Wild Plants*. N.Y.: Charles Scribner's Sons, 1977.

Bold, Harold C., *The Plant Kingdom*. Englewood Cliffs, N.J.: Prentice-Hall, 1977.

Brockman, C. Frank, *Trees of North America*. N.Y.: Golden Press, 1968.

Clark, Robert B., *Flowering Trees*. Princeton, N.J.: D. Van Nostrand, 1963.

Costello, David Francis, *The Desert World*. N.Y.: T.Y. Crowell, 1972.

Craighead, John J., Frank C. Craighead, Jr., and Ray J. Davis, *A Field Guide to Rocky Mountain Wildflowers*. Boston: Houghton Mifflin, 1963.

Crockett, Lawrence J., *Wildly Successful Plants: A Handbook of North American Weeds*. N.Y.: Macmillan, 1977.

Dietz, Marjorie J., *The Concise Encyclopedia of Favorite Wild Flowers*. Garden City, N.Y.: Doubleday, 1965.

Dodge, Natt N., *Flowers of the Southwest Deserts*. Globe, Ariz.: Southwest Parks and Monuments Association, 1973.

Dowden, Anne Ophelia, *The Blossoms on the Bough — A Book of Trees*. N.Y.: T.Y. Crowell, 1975.

Dwelley, Marilyn, *Summer and Fall Wildflowers of New England*. Camden, Maine: Down East Enterprise, 1977.

Edlin, Herbert, *The Tree Key*. N.Y.: Charles Scribner's Sons, 1978.

Eshleman, Alan, *Poisonous Plants*. Boston: Houghton Mifflin, 1977.

Evers, Robert A., and Roger P. Link, *Poisonous Plants of the Midwest and Their Effects on Livestock*. Urbana: University of Illinois Press, 1972.

Fassett, Norman C., *Spring Flora of Wisconsin*, 4th ed. Madison: University of Wisconsin Press, 1976.

Freitus, Joseph Philip, *160 Edible Plants Commonly Found in the Eastern USA*. Boston: Stone Wall Press, 1975.

Furlong, Marjory, and Virginia Pill, *Wild Edible Plants and Berries*. Healdsburg, Calif.: Naturegraph Publishers, 1974.

Gill, John D., and William M. Healy (comp.), *Shrubs and Vines for Northeastern Wildlife*. Upper Darby, Pa.: Northeastern Forest Experiment Station, 1974.

Gray, Asa, *New Manual of Botany*, 7th ed. N.Y.: American Book Co., 1908.

Greulach, Victor A. and J. Edison Adams, *Plants: An Introduction to Modern Biology*. N.Y.: John Wiley and Sons, 1967.

Grimm, William Carey, *Recognizing Native Shrubs*. Harrisburg, Pa.: The Stackpole Co., 1972.

————, *The Book of Trees*. Harrisburg, Pa.: The Stackpole Co., 1962.

Hall, Alan, *The Wild Food Trailguide*. N.Y.: Holt, Rinehart and Winston, 1973.

Hardin, James W., and Jay M. Arena, M.D., *Human Poisoning from Native and Cultivated Plants*, 2nd ed. Durham, N.C.: Duke University Press, 1974.

Harris, Ben Charles, *Eat the Weeds*. Barre, Mass.: Barre Publishers, 1968.

Hersey, Jean, *The Woman's Day Book of Wildflowers*. N.Y.: Simon and Schuster, 1976.

Hottes, Alfred Carl, *The Book of Shrubs*. N.Y.: A.T. De La Mare Co., 1942.

House, Homer D., *Wild Flowers*. N.Y.: Macmillan, 1934.

Huxley, Anthony (botannical ed.), *The Encyclopedia of the Plant Kingdom*. N.Y.: Chartwell Books, 1977.

Hylander, Clarence John, *Macmillan Wild Flower Book*. N.Y.: Macmillan, 1954.

Kelly, George W., *A Guide to the Woody Plants of Colorado*. Boulder, Colo.: Pruett Publishing Co., 1970.

Kingsbury, John M., *Deadly Harvest — A Guide to Common Poisonous Plants*. N.Y.: Holt, Rinehart and Winston, 1965.

Kirk, Donald R., *Wild Edible Plants of the Western United States*. Healdsburg, Calif.: Naturegraph Publishers, 1970.

Klein, Isabelle H., *Wild Flowers of Ohio and Adjacent States*. Cleveland: Press of Case Western Reserve University, 1970.

Klimas, John E., and James A. Cunningham, *Wildflowers of Eastern America*. N.Y.: Alfred A. Knopf, 1974.

Knap, Alyson Hart, *Wild Harvest*. Toronto: Pagurian Press, 1975.

Knutsen, Karl, *Wild Plants You Can Eat: A Guide to Identification and Preparation*. Garden City, N.Y.: Dolphin Books, 1975.

Krochmal, Arnold and Connie, *A Guide to the Medicinal Plants of the United States*. N.Y.: Quadrangle/The New York Times Book Co., 1973.

Lemmon, Robert S., and Charles C. Johnson, *Wildflowers of North America in Full Color*. Garden City, N.Y.: Hanover House, 1961.

Link, Mike, *Grazing: The Wild Eater's Food Book*. St. Paul, Minnesota: State Department of Education, 1976.

McPherson, Alan and Sue, *Wild Food Plants of Indiana and Adjacent States*. Bloomington: Indiana University Press, 1977.

Makins, F.J., *The Identification of Trees and Shrubs*. London: J.M. Dent and Sons, 1936.

Mathews, F. Schuyler, *Field Guide of American Wild Flowers*. N.Y.: Putnam, 1955.

Medsger, Oliver Perry, *Edible Wild Plants*. N.Y.: Macmillan, 1966.

Moyle, John B., and Evelyn W. Moyle, *Northland Wild Flowers: A Guide for the Minnesota Region.* Minneapolis: University of Minnesota Press, 1977.

Muenscher, Walter Conrad, *Poisonous Plants of the United States.* N.Y.: Macmillan, 1951.

Nelson, Ruth Ashton, *Handbook of Rocky Mountain Plants.* Tucson, Ariz.: Dale Stuart King, 1969.

Newcomb, Lawrence, *Newcomb's Wildflower Guide.* Boston: Little, Brown, 1977.

Niehaus, Theodore, and Charles L. Ripper. *A Field Guide to Pacific States Wildflowers.* Boston: Houghton Mifflin, 1976.

Orr, Robert T., and Margaret C. Orr, *Wildflowers of Western America.* N.Y.: Alfred A. Knopf, 1974.

Ownbey, Gerald B., *Common Wild Flowers of Minnesota.* Minneapolis: University of Minnesota Press, 1971.

Peterson, Lee, *A Field Guide to Edible Wild Plants of Eastern and Central North America.* Boston: Houghton Mifflin, 1978.

Peterson, Roger Tory, and Margaret McKenny, *A Field Guide to Wildflowers of Northeastern and North-central North America.* Boston: Houghton Mifflin, 1968.

Petrides, George A., *A Field Guide to Trees and Shrubs,* 2nd ed. Boston: Houghton Mifflin, 1972.

Platt, Rutherford, *A Pocket Guide to Trees.* N.Y.: Pocket Books, 1972.

Porter, C.L., *Taxonomy of Flowering Plants,* 2nd ed. San Francisco: W.H. Freeman and Co., 1967.

Preston, Richard J., Jr., *North American Trees,* 3rd ed. Ames: The Iowa State University Press, 1976.

Quinn, Vernon, *Shrubs in the Garden and Their Legends.* N.Y.: Frederick A. Stokes Co., 1940.

Ricciuti, Edward R., *The Devil's Garden: Facts and Folklore of Perilous Plants.* N.Y.: Walker and Co., 1978.

Rickett, Harold William, *The New Field Book of American Wild Flowers.* N.Y.: G.P. Putnam's Sons, 1963.

———, *The Odyssey Book of American Wildflowers.* N.Y.: The Odyssey Press, 1964.

Rosendahl, Carl Otto, *Trees and Shrubs of the Upper Midwest.* Minneapolis: University of Minnesota Press, 1928.

Saunders, Charles Francis, *Edible and Useful Wild Plants of the United States and Canada.* N.Y.: Dover Publications, 1948.

Simmons, Alan E., *Growing Unusual Fruit.* N.Y.: Walker and Co., 1972.

Spellenberg, Richard, *The Audubon Society Field Guide to North American Wildflowers, Western Region.* N.Y.: Alfred A. Knopf, 1979.

Spencer, Edwin Rollin, *All About Weeds.* N.Y.: Dover Publications, 1968.

Stephens, H.A., *Woody Plants of the North Central Plains.* Lawrence: The University Press of Kansas, 1974.

Sweet, Muriel, *Common Edible and Useful Plants of the East and Midwest.* Healdsburg, Calif.: Naturegraph Publishers, 1975.

Symonds, George W.D., *The Shrub Identification Book.* N.Y.: William Morrow, 1963.

Tampion, John, *Dangerous Plants.* N.Y.: Universe Books, 1977.

Tatum, Billy Joe, *Wild Foods Cookbook and Field Gude.* N.Y.: Workman, 1976.

Taylor, Kathryn S., and Stephen F. Hamblin, *Handbook of Wild Flower Cultivation.* N.Y.: Macmillan, 1963.

Thomas, John Hunter, and Dennis R. Parnell, *Native Shrubs of the Sierra Nevada.* Berkeley: University of California Press, 1974.

Thompson, Steven and Mary, *Wild Food Plants of the Sierra.* Felton, Calif.: Dragtooth Press, 1972.

Tomikel, John, *Edible Wild Plants of Eastern United States and Canada.* California, Pa.: Allegheny Press, 1976.

Van Bruggen, Theodore, *Wildflowers of the Northern Plains and Black Hills.* Interior, S.D.: Badlands Natural History Association, in cooperation with the National Park Service, 1971.

Vance, F.R., J.R. Jowsey and J.S. McLean, *Wildflowers Across the Prairies.* Saskatoon, Sask.: Western Producer Prairie Books, 1977.

Weber, William A., *Rocky Mountain Flora.* Boulder: Colorado Associated University Press, 1976.

Wherry, Edgar T., *Wild Flower Guide, Northeastern and Midland United States.* Garden City, N.Y.: Doubleday, 1948.

Williams, Kim, *Eating Wild Plants.* Missoula, Mont.: Mountain Press Publishing Co., 1977.

Wilson, Charles Morrow, *Green Treasures, Adventures in the Discovery of Edible Plants.* Philadelphia: Macrae Smith Co., 1974.

Zucker, Isabel, *Flowering Shrubs.* Princeton, N.J.: D. Van Nostrand, 1966.

INDEX OF COMMON NAMES

INDEX OF SCIENTIFIC NAMES